SOILS

William Dubbin

Published by The Natural History Museum, London

First published by The Natural History Museum,
Cromwell Road, London SW7 5BD
© The Natural History Museum, London, 2001
ISBN 0-565-09150-6

A catalogue record for this book is available from
the British Library

Edited by Karin Fancett
Designed by Mercer Design
Reproduction and printing by Craft Print, Singapore

DISTRIBUTION

Australia and New Zealand
CSIRO Publishing
PO Box 1139
Collingwood, Victoria 3066
Australia

UK and rest of the world
Plymbridge Distributors Ltd.
Plymbridge House, Estover Road
Plymouth, Devon PL6 7PY
UK

CONTENTS

PREFACE

The soils that cover the Earth's surface determine, to a large extent, the prosperity of those who depend on them for the production of food and fibre. The fertile valley soils near the Tigris and Euphrates rivers allowed the ancient civilizations of Mesopotamia to flourish. Natural flooding of these soils led to replenishment of nutrients and made possible a reliable and abundant food supply. However, degradation of these same soils through the gradual accumulation of harmful salts reduced their capacity for food production and contributed to the decline of this once prosperous region. More recent examples of soil degradation further emphasize the strong link between soil quality and human welfare. Sustained prosperity therefore depends on a thorough understanding of soils coupled with sound management practices.

This book describes the formation, properties, taxonomy and management of the world's soil resource. Each section provides an introduction to fundamental principles, and describes how these principles relate to soil properties relevant to plant growth and ecosystem function. Such a format emphasizes soils as a dynamic and exceedingly intimate blend of organic and inorganic components that occupy a central place in terrestrial ecosystems.

Also, with sections devoted to soil water, organic matter, composting, and fertility and fertilizers, the essential aspects of soil management for home gardening are made accessible to the non-expert. Whatever the reason for using this book, it is hoped that the reader will gain a greater appreciation for this indispensable natural resource.

THE AUTHOR

William Dubbin received training in soil science in Canada and the United States before joining The Natural History Museum in 1997 as a soil mineralogist in the Department of Mineralogy. He has field experience in Europe, Asia and North and Central America and has published widely on the mineralogy and chemistry of soils. His current research focuses on the associations between soil minerals and organic soil components, and the role of these associations in controlling the ultimate fate of contaminants and the maintenance of soil quality. William is a member of several soil science societies and he maintains an interest in the global issues of sustainable development and soil conservation.

SOILS – THE NUTS AND BOLTS

It is unfortunate but true that many people consider soil to be simply the weathered layer at the Earth's surface. Although there may be acknowledgement of the importance of soil for plant growth, there is frequently little awareness of either the complexity of soils or, more importantly, of the vital role of soils in maintaining ecosystem health. The following sections describe this complexity, and portray soils as dynamic natural bodies that serve essential functions in maintaining the health of terrestrial ecosystems.

WHAT IS SOIL?

Soils are naturally occurring, unconsolidated materials consisting of mineral and organic components that are capable of supporting plant growth. They cover much of the Earth's land surface, and may exceed several metres in thickness. Areas considered not to have soil are recent dumps of earth fill, bedrock, shifting sand, organic material less than 10 cm thick, and areas permanently covered by more than 1 m of water.

The functions of soil

In terrestrial ecosystems, soils have five major functions (fig 1). The first and most obvious of these is as a medium for plant growth (plate 1a). As a growing medium, soils act by providing anchorage for vegetation, supplying nutrients and water, as well as allowing for an exchange of gases between roots and the above-ground atmosphere. Second, soil serves as a habitat for a multitude of organisms.

A single handful of rich soil may contain billions of organisms representing thousands of species. How can such an abundance of organisms be supported within such a small amount of soil? This extreme diversity of organisms is due in large part to an equally diverse range of habitats that may occur within uniform-appearing soil. Third, soils are also important in the breakdown and recycling of organic materials. Without this capacity, plant and animal remains would accumulate to great depths at the soil surface. Moreover, the nutrients held within this detritus would not be available to subsequent generations of plants and animals.

FIG 1
Five functions of soil.

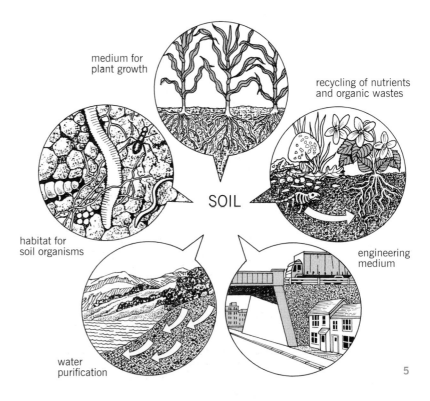

medium for plant growth

recycling of nutrients and organic wastes

habitat for soil organisms

SOIL

engineering medium

water purification

Much of the water that ends up in lakes and rivers is modified as it passes over or through soil. A fourth function of soil, therefore, is to govern the quality of water that we use domestically or for irrigation. Clean water passing through a contaminated soil may itself become polluted. On the other hand, contaminated water passing through the upper part of a soil may be cleansed of its impurities by a myriad of soil processes. Finally, soil is important as an engineering material. Most of our buildings and roads rest on soils. Understanding the properties of these soils, for example whether the soil is sandy or clayey, is important when implementing the correct construction designs.

Composition

Soils may be broadly grouped into either organic soils or mineral soils. Organic soils are those that contain more than 30%

organic matter. They occur most commonly in the cool, wet regions of the world such as northern Canada, Finland and Russia. Organic soils may be mined for their peat, which is used as a potting media, or even as a fuel in some countries. These organic resources must be used prudently to avoid unnecessary exploitation of wetland habitats. However, the vast majority of the world's soils are mineral soils, and the main focus of this book will therefore be on the properties and management of this group.

Mineral soils consist of four main components: mineral material, organic matter, air and water (fig 2). In a typical soil, the minerals and organic matter together comprise roughly 50% of the soil's volume. The remaining 50%, the pore space, is occupied by variable proportions of air and water. The amounts of air and water may change rapidly in response to variations in climate, drainage and other factors. If the pores are filled predominantly with water the soil will be waterlogged. However, if there is insufficient water in the pores a plant would suffer from drought. Optimum conditions for the growth of most plants are reached when about 25% of the soil volume is water and 25% is occupied by air.

The relative proportion of the four components of a mineral soil changes with soil depth. With increasing depth, the amount of organic matter and total pore space both decrease. Moreover, with increasing depth the pores get smaller, and they tend to be filled with water rather than air. Topsoil, the organically enriched surface layer, generally has the greatest amount of air-filled pore space, which allows for effective gas exchange with the above-ground atmosphere.

FIG 2
Soils comprise variable amounts of minerals, organic matter, air and water.

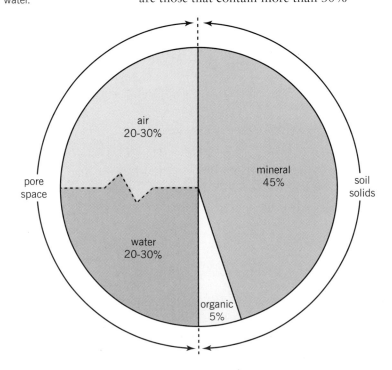

Soil as a living body

Although mineral soils consist primarily of inorganic material, they are often home to an abundant and diverse biological community. For example, 1 gram of moist topsoil may contain several billion bacteria, as well as thousands of species of fungi, roundworms and mites. Also, the roots of many plant species may extend several metres below the soil surface. These biological communities are responsible, either directly or indirectly, for virtually all natural soil reactions. Soils, as host to much of the world's biological diversity, are now known to play a major role in controlling ecosystem health.

FORMATION

The factors controlling soil formation were identified principally by two soil scientists: V.V. Dukochaev in Russia in the late 19th century, and more recently by Hans Jenny in the United States. Extensive field observations led these two researchers to conclude that the formation of all soils could be attributed to five factors. These factors are: parent material, climate, biota, topography and time.

Soil-forming factors

Parent materials are the geological or, rarely, organic precursors from which the soil formed. Examples of parent material are unconsolidated bedrock, glacial till, lacustrine deposits, alluvium, fluvial sediments and aeolian deposits. The parent material may profoundly affect the properties of the soil. For example, a soil formed from a clayey lacustrine deposit would possess mainly small pores that prevent the rapid movement of water through the profile. Such a soil may be vulnerable to flooding. In contrast, soils developed from sandy aeolian deposits would be very porous and allow easy movement of water. However, these latter soils may also be susceptible to wind erosion.

Climate refers to the temperature and precipitation regimes under which soil development occurs. Higher temperatures when accompanied with adequate water, increase both the rate of chemical reactions and the amount of plant growth. Greater precipitation increases the rate of soil formation by increasing the amount of water that percolates through the unweathered parent material. Therefore, soil forms much more quickly in the warm, humid tropics than in deserts or environments that are commonly cool or frozen.

Biota includes all living organisms, both plant and animal, associated with the developing soil. The biota contribute to soil formation mainly through the addition of organic material and the formation of soil structure. Also, vegetation stabilizes the soil surface and reduces the amount of natural soil erosion, thereby increasing the rate of soil development.

Topography refers to the form of the Earth's surface and includes elevation, slope and landscape position. Topography influences soil formation mainly through water redistribution. Lower areas in the landscape receive run-off from the higher areas and are therefore able to support a more abundant and diverse vegetation. However, if the slopes are sufficiently steep, erosion of the surface soil layers from the upper slope positions may be considerable.

Time is the fifth soil-forming factor. Although soils developed from alluvial deposits may be only several years old, and organic material may accumulate to form a darkened surface layer within a matter of decades, most soil characteristics require hundreds or thousands of years to develop. For example, a soil developed from glacial deposits on the North American plains may be 15,000 years old, whereas a highly weathered soil in Brazil may be several million years old.

It is important to keep in mind that all five soil-forming factors are interdependent and occur simultaneously. For example, across a landscape rainfall will accumulate in depressions, which increases leaching and may also give rise to a more dense vegetation. The nature of the parent material itself may vary over the landscape; certain parent materials give rise to soils that are naturally more fertile than others.

Soil-forming processes

The five factors described above interact through four soil-forming processes to produce the soil profile (fig 3). The soil profile comprises a series of layers, or horizons. These horizons vary in thickness and may have irregular boundaries, but generally they are parallel to the Earth's surface (plate 1b). The four soil-forming processes that give rise to these horizons are: (i) additions of materials such as organic matter from leaves and roots, or atmospheric dust; (ii) deletions of components, such as leaching of soluble salts or erosion of particulate matter from the soil surface; (iii) transformations, such as organic matter breakdown or mineral weathering; and (iv) translocation of material from one horizon to another.

PARENT MATERIALS

Parent material and climate are perhaps the two most important factors influencing soil properties. Parent materials are either inorganic or, less commonly, organic. Organic parent materials are deposited in wet regions,

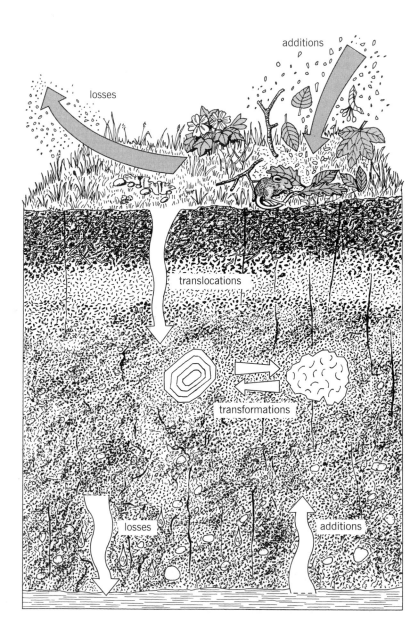

where plant material accumulates faster than it can decompose. In permanently wet and cool areas these organic deposits may accumulate to depths as great as several metres. However, the vast majority of parent materials are inorganic, and one of two types: residual or transported (fig 4).

Residual parent materials are found on all continents and consist of the great variety of unconsolidated bedrocks that occur at the Earth's surface. Igneous rocks, sandstones, mudstones and limestones are the most common types of residual parent material. The diverse composition of these materials gives rise to soils with an equally diverse range of texture, mineralogy and fertility. Transported parent materials are those that have been moved from one location to another by water, ice, wind or gravity.

Water

Water-deposited materials are either alluvial, marine or lacustrine. Alluvial sediments are those that have been deposited by rivers or streams, particularly during flooding. Soils developed from alluvial parent materials are often highly productive, as they are level, close to water and possess high fertility. Large areas of alluvial soils occur along the Amazon river in Brazil, the Nile river in Egypt and the Mississippi river in the United States. Sediments deposited in marine environments may have textures ranging from sandy to clayey. Many of these sediments also contain much sulphur, which may undergo oxidation during soil formation to produce acidity. Lacustrine (lake) deposits serve as parent material for some of the world's most fertile soils (plate 2a). Textures of these deposits are usually silty or clayey, although sandier

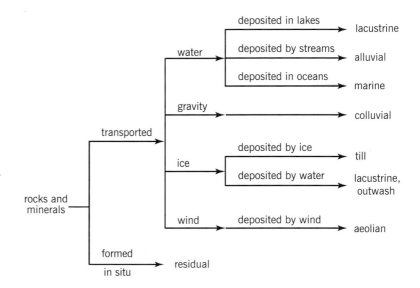

textures may occur where beaches once existed. The level topography of lacustrine soils increases their value for agricultural use, although occasional flooding and poor drainage may present a problem.

Ice

Large areas of northern Asia, northern North America, and northern and central Europe were covered by a succession of ice sheets over the last million years. Movement of the glaciers crushed the underlying bedrock and soil to produce a heterogeneous mix of material ranging in size from boulders to clay. As the glacier melted and receded, this material was deposited to form a fresh parent material, glacial till. The large volume of water gushing from the melting glacier served to sort much of this glacial till, and to form downstream an outwash plain comprised of sands and gravels. Glacial landforms are distinct from any other, and are often characterized by undulating plains and numerous ridges (plate 2b).

FIG 4
Parent materials and their mode of deposition.

Wind

Sand and silt particles from exposed and barren land surfaces may be transported by wind and deposited as a new parent material far from the source. The principal wind-deposited (aeolian) parent materials are loess and dune sand. Loess consists primarily of silt and may be carried hundreds of kilometres to its new location. Large areas of loess, occasionally reaching 50 m in depth, occur in Argentina, northern China and the central United States. Soils developed on loess generally have high fertility.

Sand may be deposited by wind in great hills and ridges called dunes, such as those in Australia and north Africa (plate 3b). With sufficient moisture, pioneering plant species may become established, thus stabilizing the dune and allowing soil formation to begin. As the sand commonly comprises quartz grains, which do not contribute nutrients, soils developed on dunes usually have very poor fertility.

Gravity

In mountainous or hilly regions, elevated and unstable material may be carried downslope by gravity and deposited as poorly sorted colluvium (plate 3a). Frost action and, occasionally, tectonic activity, assist gravity in the deposition of these frequently stony and coarse parent materials.

MINERAL WEATHERING

In response to biological, geochemical and hydrological agents acting at the Earth's surface, minerals comprising the parent material are transformed into forms that are more stable under current Earth surface conditions. Although these transformations are usually very slow, occurring over millennia, they are important for several reasons. First, during mineral weathering, nutrients such as phosphorus, potassium and sulphur are released to solution where they may be acquired by plants. Second, as the mineralogy of a soil evolves, its physical and chemical properties may change dramatically. These changes may either increase or decrease soil quality. Also, the mineralogical composition of a soil serves as an indicator of its degree of development, and may therefore reveal something of the past climate and weathering conditions to which the soil has been exposed.

Physical and chemical weathering

Minerals and rocks undergo both physical and chemical weathering. Physical weathering (disintegration) involves a reduction in the size of the rocks and minerals to ever smaller fragments, without a significant change in their composition (plates 3c, 4a). During chemical weathering (decomposition), minerals are broken down into soluble constituents that are subsequently lost to drainage water, or that reform into new and more stable minerals. Organic acids released from biological activity are especially effective at mediating the chemical weathering of minerals. In the absence of soil biota, chemical weathering would proceed at a rate probably a thousand times slower than it does currently.

Both chemical and physical weathering occur simultaneously and promote each other. The relative importance of each depends mainly on climate. Physical weathering dominates in cold, dry regions, while chemical weathering is intensified in warm and humid climates.

Mineral transformations

Relatively soluble minerals such as calcite ($CaCO_3$) and gypsum ($CaSO_4$) dissolve easily to release calcium and sulphur to solution. The weathering of these minerals is relatively simple, and does not involve the formation of new minerals. For example, a calcite grain will get smaller as it dissolves, but its composition does not change.

However, the weathering of less soluble minerals such as micas is more complex. Micas, as well as vermiculites and smectites, belong to a group of minerals called layer silicates, indicating they have a layered or leaf-like structure. Despite similar structures, these three layer silicates possess very different properties. For example, each contains a different amount of potassium, which is held with a different strength by each mineral. Over centuries and millennia, chemical weathering causes a selective removal of potassium from mica. As the potassium is selectively leached, mica gradually adopts a composition and properties like that of vermiculite (fig 5). Continued chemical weathering transforms this vermiculite into smectite.

Feldspars belong to a group of minerals called framework silicates. The weathering of feldspars is, like that of micas, highly complex. Chemical weathering of feldspars begins with the selective removal of potassium (fig 6), and leads ultimately to the formation of another layer silicate, kaolinite. The potassium released during the weathering of micas and feldspars becomes available for plant uptake. Therefore, soils that contain

FIG 5
A weathered mica grain. The expanded edge region has properties similar to vermiculite.

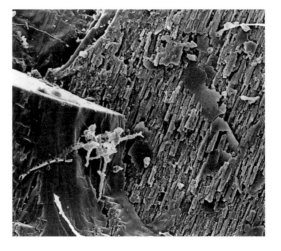

FIG 6
Chemical weathering of feldspars produces etch pits, which are visible over most of the mineral surface.

appreciable amounts of mica and feldspar are unlikely to require potassium fertilizers.

Mineralogy and soil development

Minerals vary in the ease with which they can be weathered. The suite of minerals present in the clay fraction of a soil can therefore provide an indication of the soil's degree of development. For example, gypsum, calcite and olivine weather relatively easily, and are therefore found only in the youngest soils, or perhaps older soils that occur in cold and dry climates and have consequently undergone only a limited amount of weathering. Conversely, a clay fraction dominated by kaolinite, gibbsite and goethite indicates an advanced stage of soil development. Soils consisting primarily of these latter minerals are likely to be found only in tropical climates. Based on the type of minerals that dominate the clay fraction, one can assign a development stage to any mineral soil (Table 1).

SOIL MINERALS

The suite of minerals in a soil reflects the action of soil-forming processes on the parent material. Some soils, such as those developed from dune sand, are dominated by only a few mineral types (e.g. quartz and feldspar), whereas soils developed from highly heterogeneous glacial till may consist of a great variety of minerals. Of all the minerals that may occur in soil (Table 1), those that have the greatest influence on fertility, water-holding capacity and soil structure are most commonly found within the clay fraction (i.e. particle sizes less than 0.002 mm). These clay-sized minerals are important because of their extremely small size, large surface area and electric surface charge. As most minerals possess a negative charge, plant nutrients, which are commonly present as positively charged species known as cations, can therefore be held at the mineral surface.

Minerals in the soil clay fraction

Layer silicates – The dominant minerals within the clay fraction of most soils, especially those of the temperate regions, are the layer silicates. These minerals are composed primarily of silicon, aluminium and oxygen, and are characterized by their negative charge and distinctive structures, layered much like the pages of a book. The simplest of the layer silicates is the mineral known as kaolinite (fig 7). This mineral, which appears as hexagonal crystals when viewed with an electron microscope, has only a small external surface area and small charge. Because of the small charge and surface area, this mineral has a low capacity to adsorb cations and water molecules. Consequently, kaolinite does not exhibit the stickiness, plasticity, shrinkage or swelling that characterizes many of the other layer silicate minerals. Therefore, kaolinite-rich soils provide a stable foundation for roads and buildings, although they are generally less fertile than soils dominated by layer silicates of a higher charge and surface area.

The other important clay-sized layer silicates are the smectites, vermiculites and

TABLE 1
Soil development stage

Development stage	Dominant minerals in the clay fraction
Early	Gypsum, and other soluble salts
	Calcite
	Olivine
Intermediate	Micas
	Feldspars
	Quartz
	Vermiculites
	Smectites
	Imogolite
	Allophane
Advanced	Kaolinite
	Gibbsite
	Goethite
	Haematite

fine-grained micas. Like their sand- and silt-sized counterparts, the fine-grained micas are an important source of potassium in many soils. However, the layers of this mineral are held together with such strength that they cannot expand to adsorb large amounts of water. In contrast to the non-expanding layers of fine-grained micas, the layers of vermiculite and smectite can expand when wetted to expose a large internal surface area, and in this way greatly increase the nutrient and water-holding capacity of the soil. For example, a single gram of montmorillonite, the most abundant member of the smectite group, may contain 800 m^2 of surface. The large surface area of this mineral allows for the adsorption of significant quantities of nutrients, organic molecules (including contaminants) and water, the last-mentioned of which causes considerable swelling of the clay layers. However, loss of water from smectite-rich soils, as during periods of drought, causes the layers to collapse, with the result that the soil shrinks and large cracks develop (plate 4b).

Iron and aluminium oxides – Oxide minerals of iron (e.g. goethite (fig 8), haematite) and aluminium (e.g. gibbsite) are present in small amounts in nearly all soils, and may dominate the mineralogy of highly weathered soils. Iron oxides are responsible for much of a soil's colour, with haematite providing the deep red colour of many soils in tropical regions (plate 4c). In addition, iron oxides are often used as natural pigments in paints and cosmetics. In soils, both iron and aluminium oxides commonly occur as coatings on other minerals, and in this way act as cementing agents in the formation of stable soil aggregates.

FIG 7
The layers of hexagonal kaolinite crystals are visible in this high-resolution image.

FIG 8
These crystals of goethite appear as balls of yarn in this scanning electron micrograph.

In contrast to the layer silicates, the surface charge of the oxide minerals varies with soil acidity (which is expressed as pH), and is therefore said to be a pH-dependent charge. A pH value of 7 is considered neutral; pH values lower than this are acidic, while those that are greater than 7 are alkaline. At high pH the oxides have a negative charge that is balanced by adsorbed cations, while at low pH the oxide surface has a positive charge that may hold negatively charged species known as anions. Therefore, in soils dominated by iron and aluminium oxides, pH can govern the type and amount of absorbed ions, such as nutrients and pollutants.

Other important minerals in the clay fraction – Allophane and imogolite are two poorly characterized minerals that are common in soils developed from volcanic ash, such as those that occur in many countries of the Pacific rim. As with iron and aluminium oxides, the charge on allophane and imogolite varies with pH, and in acid soils much phosphorus may be tightly bound to these minerals.

Manganese oxides, though present in only small amounts in most soils, have an importance that is much greater than their low abundance would suggest. These minerals are powerful oxidizers and play a key role in the transformations of many organic and inorganic soil components, including contaminants such as arsenic, chromium and many organic pesticides.

TEXTURE

Inorganic particles within a soil's fine earth fraction, that is all particles less than 2 mm in diameter, may be grouped into three broad sizes: sand (2.0 to 0.02 mm), silt (0.02 to 0.002 mm) and clay (less than 0.002 mm). Soil texture refers to the relative proportion of sand, silt and clay. The textural triangle in fig 9 illustrates how various proportions of these three soil separates can combine to give each of the textural classes. The most common textural classes are those of the loam group. Loam may be defined as that texture in which no single soil separate dominates. A loam texture consisting of a small excess of sand, silt or clay would be described as a sandy loam, silty clay loam or any of the other loam subclasses. Note that loam describes a textural class only, and does not, as is often incorrectly stated, refer to a fertile soil with high organic matter content.

Importance of texture
Texture is probably the single most important variable determining such fundamental soil properties as fertility, water-holding capacity and susceptibility to erosion (Table 2, page 16). Differences in many of these properties among soils can be attributed to the strong

FIG 9
Textural triangle. The three dashed lines indicate that a soil comprised of 30% clay, 30% sand and 40% silt has a clay loam texture.

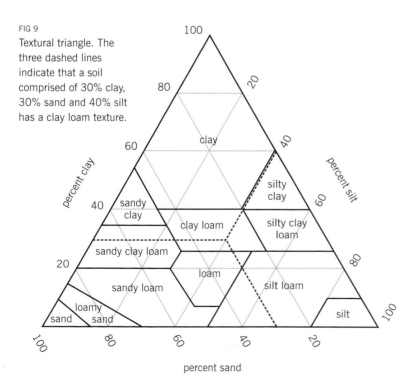

METHOD OF HAND TEXTURING

Begin with a small amount of soil (golf ball sized or larger) in one hand and knead it, while slowly adding water if necessary, until the soil has the consistency of putty. Several minutes of kneading may be required to ensure that all clumps of silt and clay have been disaggregated. It is essential that the soil have the correct water content: the soil should be uniformly moist, but there should not be any excess water.

Once the soil has reached the proper consistency, attempt to squeeze it into a ball. A soil with even a small amount of clay will form a ball easily. Next, try to form a ribbon by pressing the moist soil between the thumb and forefinger, allowing the ribbon to hang freely. The length of the ribbon will give an indication of the amount of clay. As the clay content increases, longer, shinier and more flexible ribbons will form.

To determine whether a soil is dominantly sandy or silty, add water to a small amount of soil until the soil begins to flow. A soil that is dominantly silty will feel smooth and slippery. Soils with a considerable amount of sand will feel gritty, and individual sand grains may be visible.

Follow the flow chart below to assign a textural class.

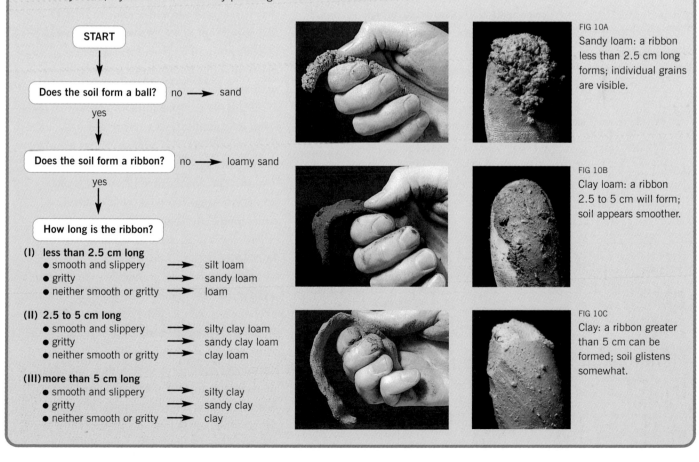

START

↓

Does the soil form a ball? no ⟶ sand

yes

↓

Does the soil form a ribbon? no ⟶ loamy sand

yes

↓

How long is the ribbon?

(I) less than 2.5 cm long
- smooth and slippery ⟶ silt loam
- gritty ⟶ sandy loam
- neither smooth or gritty ⟶ loam

(II) 2.5 to 5 cm long
- smooth and slippery ⟶ silty clay loam
- gritty ⟶ sandy clay loam
- neither smooth or gritty ⟶ clay loam

(III) more than 5 cm long
- smooth and slippery ⟶ silty clay
- gritty ⟶ sandy clay
- neither smooth or gritty ⟶ clay

FIG 10A
Sandy loam: a ribbon less than 2.5 cm long forms; individual grains are visible.

FIG 10B
Clay loam: a ribbon 2.5 to 5 cm will form; soil appears smoother.

FIG 10C
Clay: a ribbon greater than 5 cm can be formed; soil glistens somewhat.

dependence of texture on soil mineralogy. Quartz, feldspars and micas dominate the sand and coarse silt fractions, while the much more reactive oxides and clay minerals dominate the clay and fine silt fractions. The importance of texture as a fundamental soil property is further emphasized by its relative permanence. Soil texture can change only over very long periods of time through erosion, mineral weathering or translocation of particles through the soil profile.

Determination of texture

The most accurate method of determining texture is by sedimentation, which involves suspending a small amount of soil in water then allowing the individual particles to settle out. Sand settles out quickly, followed by silt and then finally clay. Although this method gives an accurate measure of sand, silt and clay abundance, it is also very time consuming and often a more rapid analysis is needed.

Textural class may be determined rapidly without equipment simply by 'feeling' the soil as illustrated in fig 10. With practice, this method can be used to accurately assign a textural class to any mineral soil.

ORGANIC MATTER

Soil organic matter consists of three broad classes of organic material: (i) living plants, animals and micro-organisms; (ii) fragments of dead plants, animals and micro-organisms; and (iii) highly decomposed and chemically variable organic compounds. The last-mentioned fraction, known as humus, makes up about 60 to 80% of the total soil organic matter. Humus is a dark-coloured collection of highly degraded organic components, such as lignins and waxes, which resist further decay. The resistance of humus to further decomposition is due not only to the highly recalcitrant nature of its constituent molecules, but also to its close association with mineral surfaces, which minimizes its exposure to micro-organisms. Like the oxide minerals, humus has a charge that is pH-dependent and, in soils of neutral or alkaline pH, humus may hold a large amount of cations. The surface area of humus is also very high, often exceeding that of the expanding clay minerals.

Functions of organic matter

The functions of soil organic matter are many and varied, influencing both plant growth and overall ecosystem health (plates 4d, 5a).
- **Nutrient supply** – soil organic matter is the primary source of nitrogen, and a major source of phosphorus and sulphur to plants. In addition, organic matter provides energy and carbon for soil organisms, and its decomposition releases beneficial vitamins and amino acids.

TABLE 2
Effect of texture on some soil properties

Property	Textural class		
	Clay	Silt	Sand
Water-holding capacity	High	Moderate	Low
Drainage rate	Slow (unless cracked)	Moderate	Fast
Water erosion susceptibility	Moderate	High	Low
Wind erosion vulnerability	Low	High	Moderate
Cohesion, stickiness, shrink–swell	High	Moderate	Low
Inherent fertility	High	Moderate	Low
Ease of pollutant leaching	Low (unless cracked)	Moderate	High
Ease of compaction	High	Moderate	Low

- **Ion exchange** – the high charge of humus allows organic matter rich soils to hold large amounts of nutrient cations in a form readily available to plants. This charge may also hold pollutants, both organic and inorganic, and prevent or retard their movement to groundwaters.
- **Structure** – organic matter serves as a glue that binds soil particles, thereby promoting aggregate formation and good soil structure.

SOILS AND THE CARBON CYCLE

All soil organic matter may be grouped into one of two fractions: active or passive. Much of the humus belongs to the passive fraction, which may persist in soils for centuries. The active fraction, however, may be degraded over months or decades. This latter fraction is very dynamic and requires careful management for soil quality to be maintained. About 4000 to 6000 kg of crop residue are required each year to maintain organic matter levels in a single hectare of land. This residue is needed to replace plant material that has been removed with the harvest, and also to replace that which has been lost to the atmosphere as carbon dioxide or methane from microbial respiration (fig 11). The release of such large amounts of these greenhouse gases to the atmosphere has important implications for global warming. Farming practices that require less tillage can be implemented to maintain near-surface soil organic residues and minimize the loss of organic carbon. Therefore, proper management of the world's soils can improve the composition of the atmosphere as well as the health of the terrestrial environment.

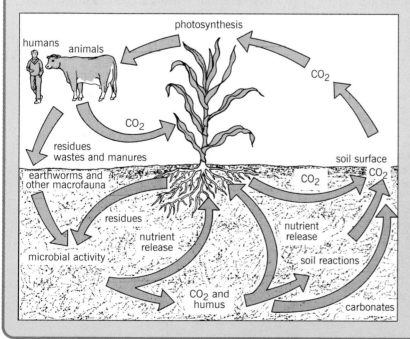

FIG 11
The carbon cycle comprises many essential reactions.

Organic matter rich soils are generally less vulnerable to erosion, and allow for easy root penetration and seedling emergence.

● **Water-holding capacity** – on a weight basis, humus can hold about five times more water than clay minerals. Organic matter also enhances water-holding efficiency indirectly, through its effect on soil structure. The more favourable structure of organic matter rich soils allows for easier water infiltration and reduced run-off.

● **Soil colour and temperature** – the dark colours characteristic of humus-rich surface horizons absorb more sunlight than lighter-coloured horizons and therefore give rise to warmer soils. Organic mulches at the soil surface buffer the soil from extremes of temperature. Gardeners frequently spread mulches over temperature-sensitive plants to provide a layer of insulation during the cold winter months. However, these mulches also retard soil warming in the spring.

Factors influencing organic matter content

The surface horizon of a well-drained mineral soil usually contains from 1 to 8% organic matter. The amount present in any given soil depends on a variety of factors. Organic matter content is greatest in cool, moist regions and lowest in warm, dry areas such as deserts. Poorly drained soils, such as those in depressions, often have a large amount of organic matter because of greater moisture from run-off and the resulting increase in plant growth. Vegetation with extensive rooting systems, such as grasses, provides more soil organic matter than do forests, which contribute organic matter mainly through above-ground leaf litter.

Cultivation greatly reduces the amount of organic matter. This reduction is the result of the removal of plant material and increased microbial decomposition caused by greater soil aeration. When a virgin soil is first cultivated, there is a rapid decline in organic matter content, especially in tropical ecosystems. However, the organic matter content eventually stabilizes at a new, lower level (fig 12).

PARTICLE ARRANGEMENT AND STRUCTURE

Structure refers to the arrangement of sand, silt, clay and organic matter into larger units. Aggregates that form naturally are called peds, whereas those that form artificially, as during ploughing or digging, are known as clods. Peds, which may range from one to several hundred millimetres in size, develop through soil-forming processes over decades and centuries as the soil matures. However,

FIG 12
Soil management practices, such as cultivation, can greatly influence organic matter content. Most changes occur within the active organic matter fraction.

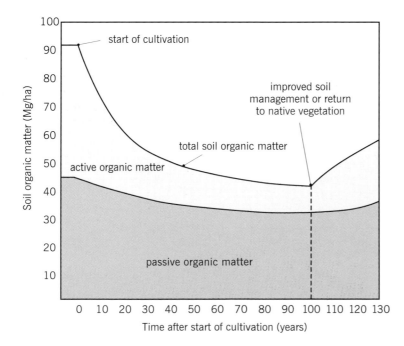

human interference can very quickly modify or destroy this structure. Soil structure greatly influences water infiltration, susceptibility to erosion, and ease of root penetration and seedling emergence. For these reasons, much effort has been directed towards understanding the factors that promote and maintain good soil structure.

Formation of soil structure

Soil peds that one may collect from a garden or a field comprise a multitude of smaller aggregates. Examination of a single ped with a hand lens will reveal the many smaller constituents which, in turn, are made up of still smaller units. Crushing a ped in your hand will similarly reveal this diverse array of ever-smaller aggregates. This structural hierarchy may be explained by first describing the interactions among the primary particles of sand, silt, clay and humus. These interactions may be classed as either biological or abiotic. Biological factors dominate in the larger aggregates and in sandy soils, while abiotic factors are most important in cementing the primary particles.

Abiotic factors – Aggregation begins with the attraction of clay particles in a process known as flocculation. This attraction occurs because cations like calcium (Ca^{2+}) and magnesium (Mg^{2+}) act as bridges between the negatively charged clay particles. Large amounts of sodium (Na^+), however, have a repulsive effect, and can cause the clay to disperse instead of flocculate. It is for this reason that soils high in sodium often have very poor structure. Humus, with its high charge, serves to bind the individual clay packets with each other and also to silt particles, thereby increasing the aggregate size (fig 13). Iron and aluminium oxides also enhance soil structure and may give rise to very stable aggregates, which become even more stable upon drying.

Shrinking and swelling, which are most pronounced in smectite-rich soils, also

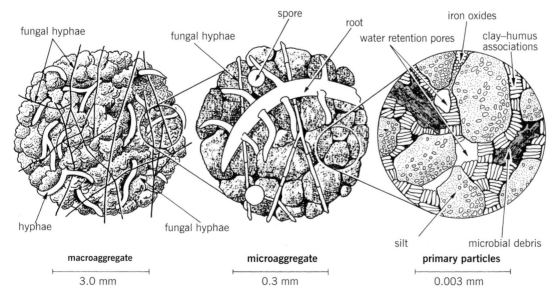

fungal hyphae

fungal hyphae

spore

root

water retention pores

iron oxides

clay–humus associations

hyphae

fungal hyphae

silt

microbial debris

macroaggregate

3.0 mm

microaggregate

0.3 mm

primary particles

0.003 mm

FIG 13
The structural hierarchy of aggregates: macro-aggregates, microaggregates and primary particles.

promote structure because the soil is forced to move along planes of weakness, and in this way help to define the ped boundaries. These processes are most common in regions with many wet/dry or freeze/thaw cycles, such as the northern plains of North America.

Biological factors – Bacteria, fungi and plant root hairs all produce polysaccharides ('sticky', sugar-like substances) that coat the surfaces of single mineral grains or clay packets. These intimate organo-mineral associations are ubiquitous in soil and help to bind microaggregates into larger units, visible with the naked eye, called macroaggregates. Microaggregates may also be bound physically by plant roots and fungal hyphae to form larger aggregates. Finally, earthworms and termites promote structure by creating channels as they burrow, and by modifying ingested soil as it passes through their gut.

Types of structure

A ped is described not only by its type (i.e. shape), but also its size (i.e. fine, medium, coarse) and grade (i.e. degree of development, such as weak, moderate or strong). There are four main types of peds, as illustrated in fig 14 and described below. Some soils, however, do not possess any aggregation or recognizable peds. These soils, such as some very young soils forming on dune sand, are considered to be structureless.

● **Spheroidal (granular)** – these are rounded peds that are common in loosely packed, organic matter rich surface soils (plate 5b). Granular peds most frequently occur in grassland soils that have been modified by earthworms.

● **Platy** – these thin, horizontal, plate-like peds are common in leached surface horizons, although they may also occur at greater depths. This platy structure is usually a product of soil-forming processes, but it may also be inherited from the parent material, or caused by compaction.

● **Blocky** – these peds are roughly cube-shaped and are usually found in subsurface horizons, where they facilitate good aeration, drainage and root penetration.

FIG 14

The four main types of structure that occur in mineral soils.

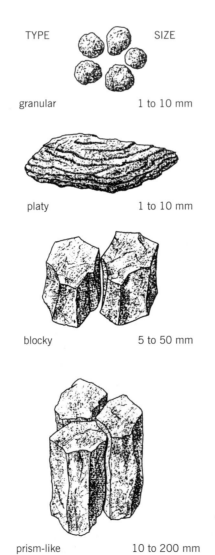

TYPE	SIZE
granular	1 to 10 mm
platy	1 to 10 mm
blocky	5 to 50 mm
prism-like	10 to 200 mm

- **Prism-like** – prism-like structures are vertically oriented columns that occur in subsurface horizons (plate 5c). The height and shape of the columns may vary. Those with rounded tops can be too dense for roots to pass through; these structures most commonly occur in soils high in sodium.

POROSITY AND DENSITY

Soil porosity and density are influenced by many factors, such as soil texture, organic matter content and structure. Porosity and density, in turn, affect water infiltration and ease of root penetration and seedling emergence. A discussion of soil porosity and density must begin with a definition of bulk density.

Bulk density

Bulk density is defined as the mass of a specific volume of dry soil. This volume includes pore space as well as solids. For this reason, bulk density will always be lower than the density of individual soil particles, which is about 2.6 g/cm^3. For example, a soil with a particle density of 2.6 g/cm^3 and with 50% pore space will have a bulk density exactly one-half the particle density (i.e. 1.3 g/cm^3). Generally, sandy soils have higher bulk densities than clayey soils. In other words, sandy soils have less pore space than clayey soils. This may at first seem difficult to believe, because sandy soils usually appear more porous than clayey soils. However, sandy soils have only a relatively few very large pores, whereas clayey soils have many extremely small pores; the total pore space is greater, therefore, in clayey soils. Organic matter decreases bulk density, mainly by increasing aggregation and promoting structure. Bulk density generally increases with soil depth because of smaller amounts of organic matter and also because of compaction from the weight of the overlying soil. As a general rule, bulk density is highest in sandy subsoils with little organic matter (1.6 to 1.8 g/cm^3), and lowest in organic matter rich, well-aggregated, clayey surface soils (1.0 to 1.2 g/cm^3).

Pore size distribution

Pore size distribution is, perhaps, more important than total pore space for determining ease of drainage, aeration and root penetration. Pores range in size from microscopic spaces (plate 5d) to large cracks. Those smaller than 0.08 mm are known as micropores, while those larger than 0.08 mm are called macropores. Macropores that have been created by living organisms, such as roots, earthworms or termites, are known as biopores. Generally, macropores dominate in sandy soils, while micropores are most abundant in fine-textured soils, especially those with little or no structure.

The size of a pore determines what function it may serve. In most soils, micropores are filled with water. These pores are often so small, however, that water movement is slow and aeration is poor. For this reason, fine-textured soils are often poorly drained and are vulnerable to flooding. The rapid movement of water and air occurs mainly through the macropores. These larger pores may also be filled with roots and a diverse assortment of tiny soil animals. In assessing soil quality, therefore, one must consider not only bulk density and total pore space, but also pore size distribution.

CAUSES, EFFECTS AND PREVENTION OF COMPACTION

Intensive use of soils through tillage, logging or recreational activities, can lead to increased bulk densities and reduced macropore space. As a result of these changes in pore space, root growth may be restricted, and water infiltration reduced, leading to greater surface run-off and a subsequent increase in water erosion.

Footpaths in public parks and forests (plate 6b) can illustrate clearly the effects of soil compaction. If foot traffic along these paths is sustained and sufficiently heavy, macropore space is reduced to the point where root growth is inhibited and plants die. There may also be a significant amount of water erosion along these trails, especially in hilly areas.

The effects of this compaction can be alleviated by loosening the soil mechanically. Unfortunately, this relief is only temporary, because such mechanical disruption destroys aggregates and results in even greater compaction in the longer term. Preventative measures are often less costly and more effective. For example, applying a thick layer of wood chips to the trail will serve to disperse foot pressure over a larger area and in this way reduce compaction. Also, compaction can be minimized by permitting traffic only when the soil is dry, and the aggregates are therefore at their strongest.

WATER IN SOIL

Water is essential for life. The growth and maintenance of plants and soil organisms, and the functioning of terrestrial ecosystems are intimately tied to the soil moisture regime. Nearly all the biological, chemical and physical processes in a soil, including those responsible for soil formation, are mediated by water. Despite the importance of water, many of the world's soils are unable to provide adequate moisture to maintain plant growth, even where a single hectare of soil may contain a million kilograms of water. How can such a situation occur? The answer is found by realizing that water in soil is quite different from water in a lake, a river or a drinking glass. Soil water is held in pores with a strength that varies with pore size. Moreover, soil water is never pure, but rather contains a diverse assortment of organic and inorganic components. Many of these dissolved substances are plant nutrients, while others are pollutants.

Water retention

Water is held in pores that range in size from large cracks to tiny interlayer spaces in clay minerals. A soil is said to be saturated when all of its pores are filled with water. Imagine that a saturated soil is placed in a flowerpot that has holes in the bottom to allow for drainage. Water will drain easily from the soil simply because of gravitational forces. When the last drop of water has fallen from the pot, the soil is said to be at field capacity. In this condition, the soil is holding as much water as it can against the force of gravity. That is, water has drained from the macropores and is present only in the micropores.

A plant growing in a soil at field capacity will initially be able to extract water easily to satisfy its needs. However, because soil water is held more tightly as the amount decreases, the plant will have progressively more difficulty in obtaining water as the soil dries. Plant roots will remove water easily first from the larger pores, then with more difficulty from the smaller pores (fig 15). Eventually, even the most drought-tolerant plants will be unable to extract sufficient moisture to survive. At this point the soil is at its permanent wilting point (plate 6c).

A clay-textured soil may contain large amounts of water at the permanent wilting point, but this water is held so tightly that it is not available to plants (fig 16). Therefore, the amount of water held between field capacity and permanent wilting point, the available water, is more important to plant growth than the total water content. Organic matter increases the amount of available water both directly, because of its greater water supplying ability, and also indirectly, through its beneficial effects on soil structure and pore space.

Water movement

The movement of water into and through a soil has implications for soil formation, plant growth, surface run-off and erosion, and also for the transport of nutrients and pollutants. Infiltration is the process whereby water from rainfall, irrigation or snowmelt enters the soil pores and becomes soil water. The rate of infiltration often varies during rainfall or irrigation. If the soil is dry initially, the macropores will be empty, thereby allowing rapid infiltration. As the macropores fill with water over time, the infiltration rate

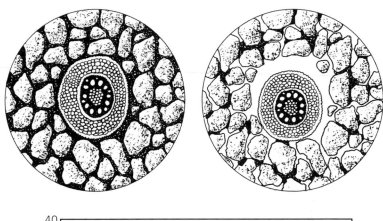

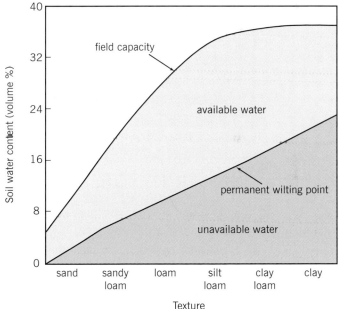

decreases and surface puddling may become apparent (plate 6d).

Once within the soil, water may move by either saturated or unsaturated flow. Saturated flow occurs when all of the pores are filled with water, whereas unsaturated flow occurs in soils whose macropores are filled with air, and therefore only the smallest pores may transmit water. In saturated flow, it is the macropores that account for most of the water movement. Therefore soils with

FIG 15 (TOP)
As soil water content decreases, less water is in contact with the root surface.

FIG 16 (BOTTOM)
Field capacity, permanent wilting point and available water vary with soil texture.

FROM DESERT TO OASIS

Temperature and sunlight levels in many arid
regions are highly favourable for plant growth.
Unfortunately, insufficient moisture in these areas
often limits crop production. Irrigation of dryland
areas, especially those with highly fertile soils
that have good drainage, can greatly increase
crop yields. There are three main methods of
irrigation: (i) sprinkler, (ii) surface (e.g. ditches
or furrows) (plate 7a), and (iii) microirrigation.
All three have benefits and drawbacks; the
method chosen will depend on a variety of
factors. An important prerequisite for all
methods, however, is an abundant supply of
clean water. Even small amounts of soluble salts
or trace metals in the irrigation water can, over
time, greatly reduce soil quality.

Many countries of arid regions, including
Egypt, Iraq, Israel, Jordan and Sudan, are highly
dependent on irrigation. The 60 million inhabitants of Egypt rely almost entirely on irrigation
for food production. The Aswan High Dam in
Egypt was constructed in the 1960s to regulate
the yearly flood of the Nile river, and also to
create a water reservoir, Lake Nasser, that could
be used to support the irrigation of vast stretches
of desert. Irrigation in Egypt has greatly
increased the agricultural productivity of a land
that was so frequently afflicted by drought. In
particular, the reliable supply of irrigation water
allowed this country to avoid the effects of the
severe droughts that occurred during the 1970s
and 1980s.

much macropore space, such as sandy soils
and those with good structure, allow for the
fastest water movement under very wet
conditions. During unsaturated flow the
water must follow a tortuous path through
a maze of capillary-sized micropores.
Consequently, the flow rate of water is
much slower in soils that are dry, or those
that are dominated by micropores, such as
clayey soils.

A third means by which water may move
in soil, as water vapour, is usually of practical
importance only in dry soils, such as those of
desert regions. Finally, it is important to
realize that in most soils water will move by
all three mechanisms simultaneously.

SOIL TAXONOMY AND GEOGRAPHY

Soils vary greatly over local, regional and global scales. The extreme diversity of soils, apparent even over small distances, reflects the range of soil-forming factors to which myriad parent materials have been exposed. The many soil types and their distribution play an important role in determining the abundance and diversity of biota within a particular region. Soil type also influences the health of ecosystems, and the ability of these systems to tolerate environmental damage, such as the accumulation of pollutants or the loss of biodiversity. Knowledge of soil taxonomy and geography can therefore provide an important foundation for understanding ecosystem functioning. A description of soil taxonomy must begin with a description of the soil profile.

THE SOIL PROFILE

The soil-forming processes described previously alter the parent material and produce a series of horizons that comprise the soil profile. The type of horizons, their thickness and sequence, help to define the soil and its properties. Often the easiest way to view a profile is to examine a road cut, many of which are sufficiently long to reveal how profiles vary across a landscape (plate 7b).

Accumulation of organic material in the top of the profile produces a dark-coloured A horizon (plate 8a). In many forest soils, leaf litter and other organic debris accumulate at the mineral surface, forming an organic (O) horizon. If leaching and weathering have been intense, a light-coloured eluvial (E) horizon may be present below the A horizon. During this leaching, clays, oxides and carbonate minerals that have been washed from the upper part of the profile accumulate in an underlying (illuvial) layer, known as the B horizon. The A, E and B horizons together comprise the solum. Below the B horizon is a layer, the C horizon, which has been relatively unaffected by soil-forming processes. The solum and C horizon are collectively called the regolith (R), which is defined as the unconsolidated mantle above bedrock. Because weathering begins at the surface, the upper horizons have been altered the most, while the horizons at the bottom of the profile are most like that of the unaltered parent material.

The letters O, A, E, B and C designate the master horizons. These upper case letters give an indication of only very general horizon properties. If a more detailed description is needed, lower case letters may be included to modify the master horizon designation. For example, Bt indicates a B horizon in which clays have accumulated (t; *ton*, German), while Bk indicates an accumulation of carbonates in the B horizon. Lower case 'b' is used to indicate buried horizons (e.g. Ab).

Soil colour and horizon delineation
Horizon boundaries may be either sharp or diffuse, and many criteria can be used to delineate one horizon from another. For example, in the field one may differentiate horizons on the basis of their colour, structure

or texture. In addition, there are a variety of laboratory analyses that can be used to distinguish horizons. The determination of structure and texture have been discussed previously. Measurement of colour, so important in soil characterization, now merits attention.

Soil colour varies not only within and between horizons, but also locally and regionally between different soils. Subtle differences in colour can reflect large and important differences in chemical and physical properties. For example, a small change in organic matter or iron oxide content would change soil colour and signify important changes in soil chemistry. Also, the presence of 'mottles' (spots, usually rust coloured) would indicate that the soil has experienced periods of poor aeration (plate 8b). A standard way to measure soil colour is by using a Munsell colour chart (plate 8c). The coloured chips on the chart are defined by three variables: hue (redness or yellowness of the colour), value (relative lightness or darkness) and chroma (purity of the colour).

SOIL TAXONOMY – AN INTRODUCTION

Classification of objects in the natural world, whether they be animals, plants, fossils or soils, is a necessary prerequisite for unambiguous and effective communication among people. Classification systems are most effective when they are systematic, detailed and understood globally. It is not sufficient, for example, for a farmer to indicate simply that he or she wishes to grow a crop in a limestone soil, a black soil or a clay soil. Such

A WINDOW TO THE PAST

Soils that have formed under previous environmental conditions, and which are now commonly buried, are known as palaeosols ('old soils') (plate 8d). Examination of palaeosols can reveal much about past climates. There is evidence of fossil soils as much as 500 million years old, although most of the palaeosols that have been studied are no more than several million years old. A striking example of how palaeosols can be used to reconstruct past climates is the impressive sequence of loess deposition followed by soil formation, spanning several million years, which has been discovered in China.

Loess is believed to be deposited during drier periods, when wind transport and deposition occur easily. Soil formation in these fresh loess deposits can begin in earnest only when the climate becomes warmer and wetter. A sequence of more than 30 palaeosols, each capped by a loess layer of variable thickness, reveals an elaborate series of wet/dry climatic cycles that continued for at least 2.5 million years in what is now north-central China. The intensity and duration of each wet period can be inferred from profile depth and the degree of mineral weathering in the corresponding palaeosol. In extreme cases, a palaeosol may possess the composition and properties of a highly weathered tropical soil, even though the current climate of the region may be cold and dry.

vague descriptions tell us little about the soil in terms of its inherent fertility, susceptibility to erosion, ability to ameliorate pollutants, or appropriate land use. In contrast, the terms Mollisol and Podzol convey the same meaning to all soil scientists around the world, and they describe a unique and specific set of properties that each soil classified as such will possess. Over the next few pages we will see how the soil profile can determine a soil's properties, and help to define how a soil is classified.

Classification systems

There are two hierarchical classification systems currently in place to classify the world's soils. Both systems have evolved over the years as our understanding of soil formation and soil properties has improved. There are likely to be additional modifications to each system as our understanding is advanced further. One of the classification systems has been developed by the United Nations Educational, Scientific and Cultural Organization (UNESCO) jointly with the Food and Agriculture Organization (FAO). This FAO–UNESCO classification scheme comprises 30 main groups, and includes an ever-increasing number of subdivisions at lower levels in the classification hierarchy. A second classification system has been developed by a team of international scientists and the US Department of Agriculture, and is currently in use to varying extents in more than 50 countries. This latter system is called *Soil Taxonomy*, and is composed of six categories arranged hierarchically, namely: order, suborder, great group, subgroup, family and series. There are 12 orders in *Soil Taxonomy*, and each of these will be

presented in the following pages to illustrate the diversity of the world's soils.

All soils can be assigned to one of the orders of *Soil Taxonomy*, based on a variety of chemical and physical properties, including the presence or absence of certain diagnostic horizons. Among the most important of these properties are mineralogy, texture, organic matter content, structure, colour, profile depth, moisture and temperature regimes, as well as a range of chemical criteria. These properties also help to define the diagnostic horizons. For example, soils developed under grassland vegetation commonly have dark, organic matter rich, high nutrient surface horizons. All soils with these distinct horizons are placed within the same order: Mollisols. On the other hand, acidic soils formed under forest vegetation may have in their B horizon large amounts of humus and iron and aluminium oxides. Soils with this diagnostic humus- and oxide-rich B horizon would be classed as Spodosols. Each of the other ten soil orders has its own diagnostic horizons and unique features.

Table 3 gives a comparison of the 12 orders of *Soil Taxonomy* with the 30 main groups of the FAO–UNESCO system, as well as a brief description of each. Andisols, Gelisols, Histosols, Spodosols and Vertisols correlate well with their FAO–UNESCO equivalents. Correlation of the other orders of *Soil Taxonomy* with their FAO–UNESCO counterparts is less precise, but the dominant features that characterize each are shared.

Gleysols and Anthrosols within the FAO–UNESCO system do not correlate directly with any single order of *Soil Taxonomy*. Gleysols are poorly drained soils that are not recognized at the order level in

TABLE 3

Comparison of *Soil Taxonomy* and FAO–UNESCO classification systems

Soil Taxonomy – order and name origin	General characteristics	FAO–UNESCO
Alfisols *alf*, nonsense symbol	Moderately leached soils; formed under forest	Albeluvisols, Lixisols, Luvisols, Planosols
Andisols *ando*, Jap., blacksoil	Formed from volcanic ejecta parent materials	Andosols
Aridisols *aridus*, L., dry	Dry soils that occur in arid regions	Calcisols, Durisols, Gypsisols, Solonchaks, Solonetz
Entisols *ent*, nonsense symbol (rec*ent*)	Young soils; very little profile development	Arenosols, Fluvisols, Leptosols, Regosols
Gelisols *gelid*, Gk., cold	Permanently frozen within 1 m of the surface	Cryosols
Histosols *histos*, Gk., tissue	Organic soils	Histosols
Inceptisols *inceptum*, L., beginning	Show only the beginnings of profile development	Cambisols, Umbrisols
Mollisols *mollis*, L., soft	Grassland soils with dark, organic-rich surface horizons	Chernozems, Kastanozems, Phaeozems
Oxisols *oxide*, Fr., oxide	Highly weathered soils; mainly found in the tropics	Ferralsols, Plinthosols
Spodosols *spodos*, Gk., wood ash	Leached, acidic forest soils; occur in temperate regions	Podzols
Ultisols *ultimus*, L., last	Strongly leached and acidic; formed under wet tropical and sub-tropical forests	Acrisols, Alisols, Nitisols
Vertisols *verto*, L., turn	Self-churning soils rich in swelling clays	Vertisols

Gleysols and Anthrosols within the FAO–UNESCO classification system do not correlate directly with any single order of *Soil Taxonomy* (see page 27)

Soil Taxonomy. Rather, they appear at lower levels in the classification as poorly drained varieties of the other orders. Anthrosols are soils that have been significantly modified by humans; they have no direct equivalent in *Soil Taxonomy*.

A number of other classification systems have been developed for use in Australia, Belgium, Canada, France, Russia and the United Kingdom. These systems are designed primarily to meet the individual needs of each country and do not attempt to provide a comprehensive description of soils globally. Correlation of soil types among these national systems is not always straightforward because terminology unique to each system makes such comparisons difficult. However, brown earth soils within the British classification system correlate broadly with Inceptisols of *Soil Taxonomy*, while rendzinas, which are shallow soils developed from limestone, would be classed primarily as Entisols.

SOIL TAXONOMY

In the following pages each of the 12 orders of *Soil Taxonomy* will be presented with reference to their unique characteristics and land use. The review will begin first with the Histosols, then the mineral soils will be discussed in the order of their increasing profile development.

Histosols

Histosols are composed largely of organic materials and occur in wetland environments, particularly those in cool regions (plates 9a, b). These soils have high water-holding capacities, but may suffer from deficiencies of various nutrients. A unique feature of Histosols is their low bulk density, which may be only 10% that of mineral soils.

With proper management, Histosols may become productive farmland, but these peat soils must first be partially or completely drained to allow for a suitably aerated rooting zone. This drainage leads unfortunately to a loss of wetland habitat, which may have once been home to a multitude of animals and plants. Drainage of Histosols leads also to rapid decomposition of organic material and the release of carbon dioxide, which adds to the problem of greenhouse warming. Furthermore, destruction of the organic material can result in significant soil losses and consequent land subsidence, particularly in warmer regions such as the Florida Everglades. A sustainable approach to Histosol use in both agricultural and forested ecosystems requires that drainage of these soils should be kept to a minimum.

Entisols

Entisols are immature mineral soils that show little profile development (plate 9d). A weak A horizon may be present in those soils where a small amount of organic matter has accumulated, but there is no evidence of a B horizon. These soils are poorly developed, either because they are very young, or because other soil-forming factors have not allowed for greater profile development. Entisols may be formed from recent deposits such as alluvium, in which case there has not been sufficient time for appreciable soil formation. In hilly areas, rates of erosion may equal rates of soil formation, thereby preventing the formation of well-developed soils (plate 9c). Climates that are very cold or dry, such as those in deserts, also restrict soil formation.

Soils that are wet through much of the year may also be poorly developed, and are therefore also classed as Entisols.

Although Entisols are poorly developed, they are extremely diverse morphologically, with an equally diverse range of agricultural productivities. Some Entisols, such as those that once formed from the alluvial floodplains of the Tigris and Euphrates rivers, may be highly fertile. Alluvium is commonly composed of nutrient-rich topsoil that has been eroded and deposited by floodwaters. Most Entisols, however, have limited productivity because of shallow profile depth, poor drainage or inadequate water supply.

Inceptisols

Inceptisols are more highly weathered than Entisols and therefore show greater profile development (plate 10a). Although Inceptisols commonly show evidence of a B horizon, with its characteristic changes in structure and colour, there has not been any appreciable translocation of clay down the profile.

The natural fertility of Inceptisols varies widely. Many have limited productivity because of either low nutrient status (e.g. low organic matter, excessively sandy) or poor climate (e.g. too cold or dry). For example, Inceptisols in the Sahelian area of west Africa are often subject to drought. This lack of moisture, coupled with overgrazing, has severely restricted the productivity of these soils. However, other Inceptisols are naturally very fertile (plate 10b) and are therefore able to support the production of high-yielding crops in some of the world's most populous regions. Most notable of these are the floodplains of the Ganges river in India, and the productive rice-growing areas of southern Asia. With proper management, however, even those Inceptisols with limited natural fertility may become highly productive.

Andisols

Andisols are formed from volcanic ash and are generally only several thousand years old. They occur most commonly near volcanoes, or perhaps downwind where the ash has accumulated to sufficient depths (plates 10c, d). They show greater development than the Entisols, but they have not weathered sufficiently to mask the properties of their unique parent materials. The recent volcanic eruptions on Montserrat in the West Indies have deposited fresh volcanic parent material from which a new area of Andisols may form.

Andisols are among the world's most fertile soils, as they possess significant amounts of organic matter and high nutrient and water-holding capacities. Unfortunately, they often bind phosphorus with such strength that deficiencies of this nutrient may occur in those soils that are not managed properly. Also, because Andisols commonly occur on rather steep slopes, they may be susceptible to erosion if precautions are not taken.

Gelisols

Gelisols are soils that have a permanently frozen layer within a metre of the surface. Because of the cold climate in which these soils occur, there has been little profile development. A property unique to Gelisols is cryoturbation, which refers to the physical churning of the soil caused by freezing and thawing of water. This frost action may be visible in modern Gelisols, as well as those that were formed during previous colder climates. In the latter case, the convoluted horizons characteristic of cryoturbation remain as relics of a previous ice age (plate 11b).

Although human populations are sparse in areas dominated by Gelisols, these soils provide migratory birds and caribou with, respectively, essential nesting sites and tundra vegetation during warmer seasons. However, the permafrost of Gelisols presents engineers with considerable challenges when building roads and other structures (plate 11a).

Aridisols

Aridisols are soils of desert regions that have light-coloured surface horizons with little organic matter (plate 12a). Because of low precipitation, these soils show little development, although soluble minerals such as calcite and gypsum may have been leached from the upper part of the profile and deposited at a lower level.

Aridisols occur in areas that, without irrigation, are generally too dry to support crops (plate 12b). Irrigation can greatly increase the productivity of these soils, although careful management is required to prevent the accumulation of salts and sodium, both of which can be toxic to plants. Many Aridisols, such as those in the deserts of the western United States, may be used for low-intensity grazing, although the consequent reduction in vegetative cover on these sparsely covered soils increases their vulnerability to erosion. Many Aridisols in sub-Saharan Africa have been degraded by such erosion.

Vertisols

Vertisols consist of at least 30% swelling clay and they therefore undergo intense shrinkage and swelling during periods of drying and wetting. This repeated movement of soil along planes of weakness gives rise to peds with shiny surfaces known as slickensides (plate 11d). A Vertisol profile often shows little horizon differentiation, and is often darkly coloured, even when organic matter contents are low.

Vertisols are often difficult to cultivate because of their large amount of swelling clay (plate 11c). They become very sticky when wet and very hard when dry, necessitating the use of powerful tractors if large areas are to be tilled. These adverse physical characteristics present special problems in India and the Sudan, where Vertisols are common, but often only slow-moving animals are available to assist with tillage. The shrink–swell properties of these soils also pose considerable engineering problems. Highways, railways and buildings constructed on these soils must often incorporate expensive construction designs to prevent their cracking or deformation. Despite the considerable management problems associated with Vertisols, these soils are often very fertile and, if managed properly, can be productive owing to their high clay content.

Mollisols

Mollisols develop under grassland vegetation where the extensive rooting systems of the grasses produce soil surface horizons rich in organic matter and calcium (plates 13a, b). The thickness of these humus-rich A horizons may approach one metre in humid regions. The surface horizons of Mollisols usually have a granular structure, while in the B horizon favourable blocky structures commonly occur.

High organic matter content and favourable structure make Mollisols some of the most productive soils in the world. These soils are used widely to produce many crops, especially wheat, and for this reason are an important component of the world's 'bread basket'. When these soils are first ploughed, as they were more than a century ago on the North American plains, nutrients are released from the surface horizon in quantities sufficient to give high crop yields. However, occasional drought limits the productivity of some Mollisols. The lack of moisture may also increase their susceptibility to wind erosion, especially in those soils that have developed from loess.

Alfisols

Alfisols commonly develop under deciduous forests in humid temperate regions (plates 14a, b). Because they have developed under climates wetter than those under which Mollisols have formed, they show greater leaching of salts and other soluble constituents. Alfisols are characterized by a B horizon in which there has been an accumulation of clay. If this clay accumulation has been significant, an E horizon may be apparent above the B horizon, and within the B horizon shiny clay skins are usually visible on the ped surfaces.

Because of their high nutrient status, favourable textures and occurrence in suitable climates, Alfisols are useful for the production of many crops and hardwood forests. Throughout history many prosperous societies have relied on Alfisols for the production of much of their food and fibre. However, these soils are vulnerable to both wind and water erosion if their organic matter rich surface layer is lost. Also, slow water infiltration into the clay-rich B horizon may result in the waterlogging of some Alfisols during periods of high rainfall.

Spodosols

Spodosols develop in cool, wet climates, usually under coniferous forests (plates 13c, d). The sandy, acidic parent materials from which these soils have formed allow for rapid water infiltration and the formation of a highly leached E horizon. The underlying B horizon is rich in humus and oxides that have leached from the upper part of the profile. Spodosols, with their white E horizons and black and reddish-brown coloured B horizons, are one of the most distinctive of all the orders within *Soil Taxonomy*.

Low natural fertility and limited water-holding capacity make Spodosols unsuitable for most agricultural uses. The large amount of fertilizers and lime required to support the growth of annual crops on these soils often proves too costly. Consequently, most Spodosols remain under their natural vegetation, where they may find use in forestry.

Ultisols

Ultisols are highly weathered soils that have formed over long periods of time in tropical or subtropical climates under forest or savanna vegetation (plate 15a). Long periods of intense leaching have produced acidic soils with clay-rich B horizons composed mainly of kaolinite, with smaller amounts of iron oxides, montmorillonite and fine-grained micas. An Ultisol profile is usually yellow or red, reflecting the accumulation of iron oxides (plate 15b).

PLATE 1A
Soils are used by humans in all parts of the world to produce food and fibre.

PLATE 1B (FAR LEFT)
Soil horizons may vary in thickness but are parallel to the Earth s surface.

PLATE 1C (LEFT)
Deep and highly weathered soils frequently occur under warm and humid climates, such as this giant Podzol from south-east Asia.

PLATE 2A
Lake sediments, comprised mainly of silts and clays, become lacustrine parent materials from which fertile soils may form.

PLATE 2B
Ridges of glacial till deposited by a receding glacier.

PLATE 3A
Colluvium serves as parent material for many soils in hilly and mountainous regions. Exposed bedrock (background) and colluvium frequently occur together.

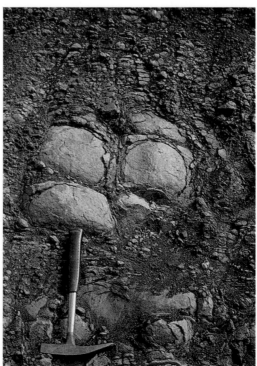

PLATE 3B (FAR LEFT)
Wind may transport and deposit vast amounts of silt and sand.

PLATE 3C (LEFT)
Physical weathering of these rocks has resulted in characteristic 'onion skin' weathering.

PLATE 4A (RIGHT)
Physical weathering of bedrock during soil formation: disintegration is most intense near the surface.

PLATE 4B (FAR RIGHT)
Smectite-rich soils frequently develop large cracks during periods of drought.

PLATE 4C (RIGHT)
Iron oxides, particularly haematite, are responsible for the deep red colour of many highly weathered soils in tropical regions.

PLATE 4D (FAR RIGHT)
The leaf litter on this forest floor is an important component of soil organic matter.

PLATE 5A (FAR LEFT)
Organic matter gives a dark colour to surface horizons.

PLATE 5B (LEFT)
Granular structure promotes water infiltration, and ease of root penetration and seedling emergence.

PLATE 5C (FAR LEFT)
Prism-like structures commonly occur in subsurface horizons. If these peds are very dense, root penetration may be inhibited.

PLATE 5D (LEFT)
Thin-section showing individual sand and silt grains, as well as clay coatings (brown). Black areas indicate pore space.

PLATE 6A (RIGHT)
Dense surface crusts inhibit plant growth and often require mechanical disruption.

PLATE 6B (FAR RIGHT)
Unless proper measures are implemented, heavily used trails in recreational areas can cause soil compaction and increased water run-off, leading to soil erosion.

PLATE 6C (RIGHT)
In the absence of sufficient available water, these water-stressed plants will die once they reach the permanent wilting point.

PLATE 6D (FAR RIGHT)
A saturated soil, or a soil with a slow rate of water infiltration, can quickly become ponded during sustained rainfall or irrigation.

PLATE 7A
Surface irrigation, using ditches and furrows, can provide an effective means of providing water to crops in dryland areas.

PLATE 7B
The dark, organic matter rich A horizon shown in this road cut from South Africa is thickest at the left, most likely as a result of increased water from run-off and a consequent increase in plant growth.

PLATE 8A (RIGHT)
The A, E and B horizons can be seen clearly in this forest soil.

PLATE 8B (FAR RIGHT)
Mottles are reddish spots within a blue–grey matrix that indicate the soil has experienced periods of poor aeration.

PLATE 8C (RIGHT)
The coloured chips of a Munsell colour chart allow for a standardized measurement of soil colour.

PLATE 8D (FAR RIGHT)
A well-preserved sequence of palaeosols can be used to reconstruct past climates.

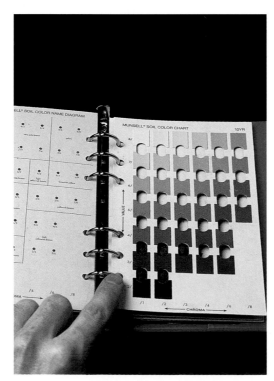

PLATE 9A (FAR LEFT)
Histosols commonly
occur in regions that
are wet and cool,
such as this area in
Newfoundland Canada.

PLATE 9B (LEFT)
Histosols often show
poor horizon delineation.

PLATE 9C (FAR LEFT)
Entisols frequently
dominate rugged and
eroded landscapes.

PLATE 9D (LEFT)
These Entisols from Iraq
form a thin veneer over
bedrock.

PLATE 10A (RIGHT)
The thick, organic matter rich A horizon of this Inceptisol indicates a relatively high fertility.

PLATE 10B (FAR RIGHT)
Fertile Inceptisols may be found on these elevated alluvial plains near the Awatere River, New Zealand.

PLATE 10C (RIGHT)
Andisols, like this soil from Japan, frequently occur over buried profiles.

PLATE 10D (FAR RIGHT)
Volcanic eruptions release ash and other debris, which serve as parent material for all Andisols.

PLATE 11A (FAR LEFT)
These distinctive polygons, characteristic of many regions dominated by Gelisols, are created by frost heaving.

PLATE 11B (LEFT).
The convoluted horizons of these soils near Paris, France, remain as relics of Gelisols that existed during a colder climate.

PLATE 11C (FAR LEFT)
Vertisols occur on flat landscapes and frequently have heavily cracked surfaces (foreground).

PLATE 11D (LEFT)
Shiny ped surfaces are characteristic of Vertisols.

PLATE 12A
Large accumulations of calcite and gypsum are visible as white deposits in the lower part of this Aridisol from New Mexico, United States.

PLATE 12B
Low precipitation in desert regions restricts both vegetative growth and soil organic matter.

PLATE 13A (FAR LEFT)
Mollisols develop under grassland vegetation, similar to this region of Saskatchewan, Canada.

PLATE 13B (LEFT)
The A, B and C horizons of this Mollisol are clearly visible.

PLATE 13C (FAR LEFT)
Spodosols are frequently associated with coniferous forests.

PLATE 13D (LEFT)
Spodosols are among the most distinctive of soils, with well-defined A, E, B and C horizons.

PLATE 14A
The leached E horizon of this Alfisol is visible as a grey layer in the upper part of the profile.

PLATE 14B
Alfisols typically develop under deciduous forests.

PLATE 15A (FAR LEFT)
Ultisols support the growth of this banana plantation in Rwanda.

PLATE 15B (LEFT)
The horizons of an Ultisol profile are visible in this soil pit.

PLATE 15C (FAR LEFT)
Oxisols form on stable landforms in the warm, humid tropics, such as this area of Kenya.

PLATE 15D (LEFT)
Oxisol horizons are poorly delineated and are composed largely of iron and aluminium oxides, conferring a deep red colour to these soils.

PLATE 16
The global distribution of
soils mainly reflects
variable climate and
parent material.

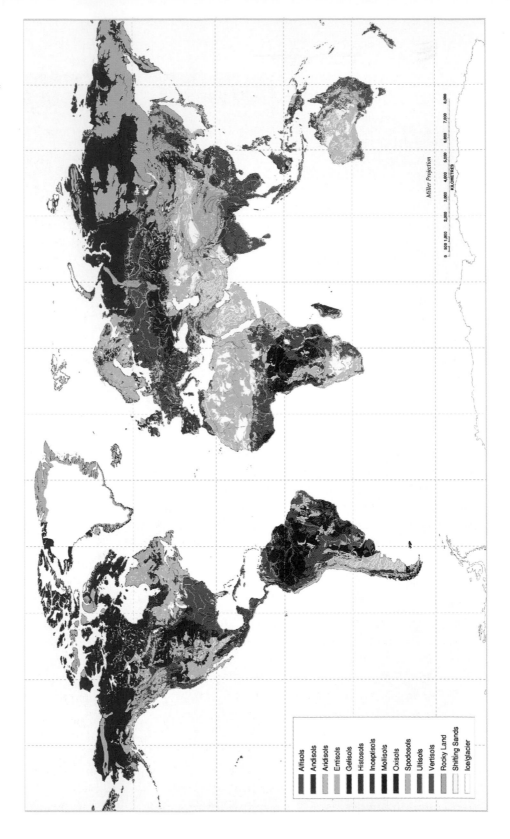

Intense weathering during Ultisol formation has leached many nutrients below the rooting zone. The addition of fertilizers and lime to these soils can, however, increase their fertility to levels that rival those of the Alfisols or Mollisols. Also, most Ultisols occur in regions with abundant moisture and long growing seasons. With proper management, Ultisols can therefore be among the world's most productive soils.

Oxisols

Oxisols are the most highly weathered soils and typically occur on old, stable landforms in the humid tropics (plate 15c). They are composed of only the most resistant minerals, principally iron and aluminium oxides, kaolinite and quartz. These soils, which may be more than 10 m deep, show little horizon differentiation and, owing to their large haematite content, are often deep red in colour (plate 15d).

When Oxisols are first cleared and cultivated, their productivity may be quite high. Their organic matter is broken down rapidly in the warm, moist climate, thereby releasing a flush of nutrients. However, once the organic matter has been largely depleted, nutrient levels decrease rapidly because the highly weathered minerals in these soils cannot release additional nutrients in sufficient quantities. Consequently, Oxisols may quickly become unproductive if their native vegetation is removed and they are not subsequently managed properly. Perhaps the most appropriate use for Oxisols is in the production of various tree crops, as well as for the support of our tropical rainforests. With proper management and adequate inputs of fertilizers, especially phosphorus,

these soils may support the production of, for example, bananas, coffee and pineapples.

SOIL GEOGRAPHY

Approximately 13% of the Earth's ice-free land surface is covered by rock or shifting sand; these areas are considered to not have soil. Within this ice-free land surface, Entisols are the most abundant soils, covering 16% of the area. They are found globally and occur in many climatic regions, but are most abundant in the Sahara desert, central Australia, and the upland regions of Iran and Pakistan (plate 16). Aridisols are the next most abundant soils, occupying 12% of the Earth's ice-free land surface. The largest areas of Aridisols occur in southern Australia and the Gobi desert in China, as well as the deserts of the western United States and southern Argentina.

Alfisols and Inceptisols each occupy 10% of the global ice-free land area. Alfisols are found in cool, moist climates, with the largest area occurring in the Baltic States and western Russia. Inceptisols, however, are found in most climatic regions and on all continents, where they are often the dominant soils of mountainous areas. Gelisols (9%) occur mainly in the far north, although smaller areas may be found at higher elevations in less northerly latitudes. The productivity of these soils is limited not only by their permafrost but also by the short growing season in the far north.

Oxisols (8%) generally occur in the tropics, most notably in Brazil and central Africa, where weathering conditions are most intense. Ultisols (8%) also occur in the warm, humid areas of the lower latitudes, often in association with Oxisols. Mollisols are only slightly less abundant, comprising 7% of the land area. Because of their high fertility and suitability for agriculture, few Mollisols remain uncultivated. They cover vast areas of the Ukraine, Russia and Kazakhstan, as well as the interior plains of North America.

Only 3% of all soils are classed as Spodosols, yet these soils support many of the important coniferous forests in northern Europe and north-eastern North America. By virtue of their low fertility, however, few Spodosols are used for intensive agriculture. Although Vertisols are also limited in their extent, comprising only 2% of the world's soils, important areas occur in Australia, India and the Sudan. Those in India are especially significant, as they are used to grow crops that feed one of the world's most populous regions.

Histosols (1%) occur mainly in the wet, cool regions of Finland and northern Canada. Despite their low abundance, these soils form an integral part of many environmentally important wetland ecosystems. Although Andisols are the least abundant of all soils, occupying less than 1% of the Earth's ice-free surface, significant areas occur in Japan, Chile and Mexico. Their low abundance does not, therefore, diminish their importance, as they are used successfully for intensive agriculture in many populous regions of the Pacific rim.

SOIL BIOLOGY

Soils are teeming with living organisms and are host to much of the world's biological diversity. A single kilogram of rich topsoil may contain 1 billion fungi, 10 billion actinomycetes, and more than 500 billion bacteria. Even the population of fauna can exceed 500 million within a kilogram of soil. Organisms are present in soil not only in vast numbers but also with a diversity that is probably greater than for any other ecosystem. For example, many scientists believe that within the Amazon rainforest there are more species of organisms in the soil than within the forest itself.

Soils are an important source of much of the world's species diversity. This diversity of species gives rise in healthy soils to functional diversity: the ability of a soil to carry out a range of processes using many different energy sources. This functional diversity, in turn, leads to functional redundancy, which is a property of those soils in which any given biochemical process could be carried out by more than one species. Such redundancy is important because it confers to a soil both resilience and stability. A soil is said to be resilient if it can return quickly to healthy functioning following a significant disturbance. Stability, on the other hand, refers to the capacity of a soil to continue essential functions such as degradation of organic residues and cycling of nutrients, even in the presence of ongoing environmental disruption. It is for these reasons that the abundance and diversity of organisms in a soil are often used as diagnostic indicators of soil quality.

ORGANISMS AND THEIR FUNCTION

Soil organisms consist of animals, plants, fungi and micro-organisms. All of these interact through a complex network of energy and nutrient transfers known as a food web. Organisms within this web are assigned to broad categories based on their primary food source. Organisms that feed directly on living plant residues are known as herbivores, and they are the primary consumers. Parasites and predators that feed on primary consumers are known as secondary consumers (e.g. protozoa, nematodes). Organisms that prey on the secondary consumers are called tertiary consumers (e.g. ants, spiders). Still other organisms feed on dead organic matter; they are known as decomposers. Note, however, that the interactions within this web are complex and the distinctions among the four groups are not always clear.

Animals (fauna)
Soil fauna greater than 1 mm in width, such as earthworms, termites, ants, beetles and gophers, are known as macrofauna. Of all the soil macrofauna, earthworms are probably the most important group for improving soil quality (fig 17). As earthworms burrow through the soil, ingesting each day an amount of soil nearly ten times their own weight, they create an intricate and extensive network of biopores that improve both aeration and drainage. The soil ingested by the earthworms is eventually expelled as

FIG 17
Earthworms promote soil structure as they create burrows and degrade organic debris.

FIG 18
Termite mounds are a distinctive feature of many landscapes in tropical and subtropical regions.

FIG 19
Dung beetles influence nutrient cycles as they transport and store balls of animal dung.

nutrient-rich casts. These casts are often deposited within the burrows, where plant roots also frequently occur, and in this way help to facilitate the exchange of nutrients. Several thousand species of earthworms, most ranging from 1 mm to 1 m in length, have been identified globally. They are most abundant in moist, but well-drained, organic matter rich soils. Consequently, earthworms occur most frequently in Mollisols and other soils rich in organic material.

Termites, which occur mainly in tropical and subtropical regions, influence soil properties on a scale similar to, or perhaps greater than, that of the earthworms. Termites build nests by gathering soil from lower levels, which is then transported upwards to construct great chambered mounds that may extend more than 5 m above the soil surface (fig 18). Construction of these mounds also involves the creation of a network of robust underground macrochannels that may extend outwards more than 10 m from the mound. When the mound is removed, these channels remain to facilitate rapid drainage of waters that may otherwise collect at the soil surface.

Other important soil macrofauna include the ants, beetles, and vertebrates such as gophers. Ants are important because they are key tertiary consumers and, like termites, they may also move large amounts of soil as they build their nests. Dung beetles influence nutrient cycling by storing balls of nutrient-rich animal dung in shallow, subsurface channels (fig 19). Burrowing animals such as gophers aerate and mix the soil. The activities of these animals are often evident by the presence of crotovinas: burrows in one horizon filled with soil from another.

Fauna less than 1 mm in size, such as protozoa and nematodes, are known as microfauna. Protozoa are a morphologically diverse group of single-celled organisms that prey on bacteria, fungi and other microbes. More than a thousand species of protozoa have been isolated from soil and these can be grouped into three classes according to their method of movement in water. Amoebae move by means of pseudopodia, ciliates wave their hair-like cilia (fig 20), and flagellates move by using long, whip-like appendages called flagella. The principal means through which protozoa influence the soil ecosystem is by controlling the population of micro-organisms, and in this way altering the various nutrient cycles. Although soil-dwelling protozoa cause few plant diseases, waterborne protozoa are the cause of several important human diseases, most notably cryptosporidiosis and amoebic dysentery.

Nematodes are microscopic worms that occur in nearly all soils (fig 21). Most nematodes survive by feeding on bacteria and fungi, while the more harmful varieties obtain nutrients by infesting plant roots, and in so doing allow for the easy entry of other pathogens. Corn (maize), soybeans and sugar beets are particularly susceptible to nematode infestation. However, plants such as canola and marigolds produce root exudates that keep nematode populations in check.

Plants (flora)

Most living plant roots are classed as macroflora (>1 mm), although the smallest root hairs may be considered microflora (<1 mm). A single plant may have, within the top metre of a soil, roots with a combined length exceeding 600 km. Roots release many substances that increase nutrient availability, thereby enhancing plant growth and altering the environment for nearby micro-organisms. These microbes, in turn, influence the chemical and biological environment of the root zone. This region immediately around the root that has been significantly altered by the root and its exudates is called the rhizosphere.

FIG 20 (LEFT)
This ciliate, *Colpoda steinii*, is but one of more than 1000 known species of soil protozoa.

FIG 21 (RIGHT)
Nematodes may feed on plants, animals, fungi and bacteria.

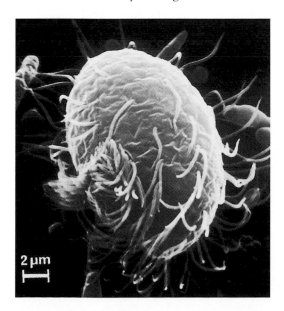

2 μm

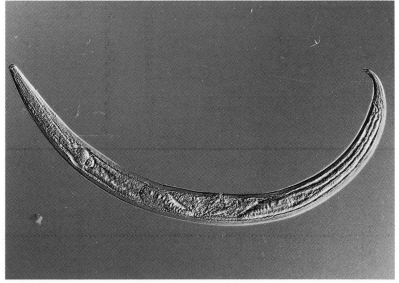

FIG 22
The fruiting body of the *Amanita* sp. This filamentous fungi grows in close association with plant roots.

Algae, an important member of the microflora, are photosynthetic organisms that are sometimes visible near the surface of moist soils. Although algae are not important as decomposers of organic residues, they may add to a soil significant amounts of nitrogen, as well as polysaccharides that promote the formation of structure.

Fungi

The fungi encompass a diverse assortment of yeasts, which are single-celled, and multicellular filamentous moulds and mushrooms (fig 22). Filamentous fungi are composed of long, branching chains of cells called hyphae. These hyphae may be interwoven to form thread-like mycelia that are often visible in leaf litter as thin white strands. The filamentous fungi effectively decompose virtually all forms of organic material: tree bark, leaf litter, decaying animal remains, and also food, as evidenced by the frequent occurrence of these organisms on rotting bread and cheeses. Some filamentous fungi also have highly important symbiotic associations with plant roots forming mycorrhizae, which means 'fungus root'. Because of the many benefits conferred to the plant by mycorrhizae (e.g. increased nutrient availability), these associations occur in nearly all soils. In contrast to the benefits of mycorrhizae, the presence of mushroom fungi in lawns can be a nuisance to many gardeners. The fruiting bodies of these fungi often form unsightly and ever-expanding 'fairy rings' as their associated underground hyphae grow outwards into uncolonized soil.

Micro-organisms

Bacteria are the most abundant and diverse of all soil organisms. Because of their functional diversity, they are involved in virtually every metabolic process within a soil. A particularly important bacterially mediated process is the transformation of nitrogen from one form to another. This essential nutrient, if not in the correct form, may be toxic to plants or leached easily from the soil profile. Soil bacteria are important also in forming mutually beneficial (i.e. symbiotic) relationships with legumes such as alfalfa, beans and peas. In these plants, bacteria of the genus *Rhizobium* form specialized root nodules that allow for the fixation of atmospheric nitrogen, which can then be acquired by the plant. In return, the bacteria receive from the plant various carbohydrates for energy.

Actinomycetes are related to both bacteria and fungi and appear as highly branched filamentous structures. They survive by decomposing organic material and therefore they are most abundant in organic matter rich soils and compost heaps. Actinomycetes are important not only because they effectively decompose even the most resistant organic constituents, such as cellulose, but also because they produce many important antibiotics. Since the 1950s more than 500 antibiotics have been isolated from actinomycetes; the most notable of these are actinomycin and streptomycin. Actinomycetes also produce a variety of compounds, collectively called geosmins, which are responsible for the rich, earthy aroma often encountered when digging garden soil.

BENEFICIAL USES OF SOIL MICRO-ORGANISMS

Soil organisms are important in the production of antibiotics, the remediation of polluted soils, and in maintaining soil fertility. Another beneficial application of micro-organisms involves the use of *Bacillus thuringiensis* (Bt) in the biological control of herbivorous insects. This naturally occurring bacterium produces a toxin that, when ingested by the larvae of these pests, causes irreparable damage to their gut and in this way effectively controls their population. Large amounts of the Bt toxin have been isolated from bacterial cultures and incorporated into various formulations for spraying on trees and crops, thus saving farmers from significant losses.

Recent advances in molecular biology have allowed this pesticide to be used with greater specificity. For example, the gene that produces the Bt toxin has been isolated from *B. thuringiensis* and transferred into another bacterium, one that is abundant in the rhizosphere of many plants. Seeds or roots of vulnerable plants, such as corn, can then be inoculated with this genetically engineered bacterium as a defence against herbivorous pests. Of course, the benefits of such engineering must be balanced against potential dangers. The introduction of genetically engineered micro-organisms is not without risks, as the modified organism may alter the soil ecosystem in ways that are not always easy to predict. It is therefore prudent to test each new organism thoroughly before it is introduced to the environment.

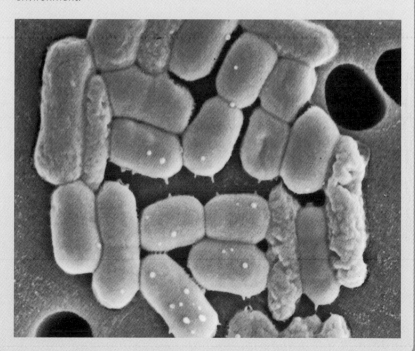

FIG 23
Bacteria are the most diverse and ubiquitous soil organisms.

SOIL FERTILITY

Plants require water, sunlight, heat and nutrients to survive and grow. Optimum plant growth is achieved when these factors are supplied in the correct amount and proportion. Soil fertility is that branch of soil science concerned with optimizing the supply and availability of plant nutrients. Of all the disciplines within soil science, fertility is the one most directly concerned with plant growth, and devising ways to improve a soil's fertility to increase crop yields or grow better gardens.

THE FUNDAMENTALS

Several concepts are fundamental to soil fertility, and these must be considered when making decisions concerning practical soil nutrient management. For example, what are the benefits and limitations of using organic versus inorganic fertilizers? And also, how should the nutrient status of a soil be modified to ensure optimal plant growth?

The basic principles of soil fertility, described below, apply to all soil–plant relationships, whether one wishes to grow wheat, roses, pineapples, tomatoes, redwood trees or any other plant.

Essential elements

Eighteen elements are essential for plant growth (Table 4). All of these elements must be available in sufficient quantities if normal plant growth and reproduction are to occur. Nine of the essential elements are required by plants in relatively large amounts and are called macronutrients. The remaining nine elements are needed in much smaller quantities and are therefore known as micronutrients. Several other elements may also be taken up by plants, but they are believed to be not essential for growth.

Plants obtain carbon, hydrogen and oxygen from air and water; the other elements must be acquired from the soil minerals and organic matter. Deficiencies of micronutrients are rare, as most soils are able to supply these nutrients in the small amounts required for normal plant growth. The much greater demand for macronutrients, however, often results in deficiencies of these elements. Much of soil fertility is therefore concerned with the macronutrients that are derived from the soil solids: nitrogen, phosphorus, potassium, sulphur, calcium and magnesium.

Total versus bioavailable nutrients

If one were to analyse a soil for both its total phosphorus content as well as that which is

TABLE 4
Elements essential
for plant growth

Macronutrients	Micronutrients
Carbon (C)	Iron (Fe)
Hydrogen (H)	Manganese (Mn)
Oxygen (O)	Boron (B)
Nitrogen (N)	Zinc (Zn)
Phosphorus (P)	Copper (Cu)
Potassium (K)	Chlorine (Cl)
Sulphur (S)	Molybdenum (Mo)
Calcium (Ca)	Cobalt (Co)
Magnesium (Mg)	Nickel (Ni)

available to plants, it would be apparent that only a small fraction of the total soil phosphorus is bioavailable. This general rule applies also to the other nutrients. Why is this so? And what does this mean for a plant growing in this soil?

Plant roots cannot obtain nutrients by ingesting whole soil particles. Rather, the nutrient elements within these particles must first be dissolved in the soil water, where they can then be absorbed by plant roots and micro-organisms. Nutrient elements in the soil solution are therefore said to be plant-available (fig 24). The amount of nutrient in the soil solution is usually less than 1% of the total nutrient content of a soil. However, nutrients within the soil solution are subject to considerable changes in concentration over short periods of time, because of uptake by plants or losses by leaching.

Because nutrients in solution exist as charged species, either cations or anions, they may be held by the charged surfaces of minerals and organic matter. For example, large amounts of potassium cations can be bound to the negatively charged surfaces of clay minerals. In fact, there may be more than ten times as much potassium held at these surfaces as there is in solution. During a very important process known as cation exchange, various cations in solution exchange with nutrient cations (e.g. potassium) held at mineral and organic surfaces, thereby replenishing the soil solution with nutrients. Nutrients held at surfaces in such a loose manner are said to be exchangeable and, because they can move readily between the surface and solution, comprise the labile pool of nutrients. Those nutrients held within very soluble minerals and rapidly decomposable

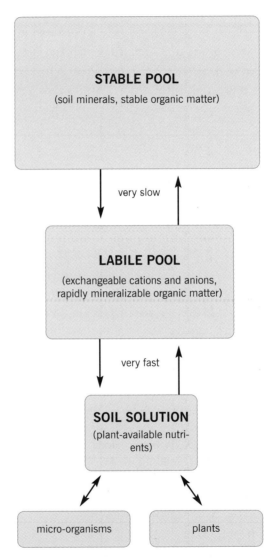

FIG 24
Schematic illustration showing the composition and relationships among the three nutrient pools.

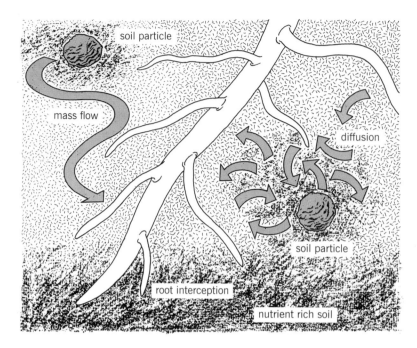

FIG 25
Nutrients are replenished
at the root surface
by three principal
mechanisms.

and therefore practically useless as a nutrient source. From the perspective of a growing plant, only the amount of available nutrient is of importance.

Nutrient uptake

Nutrients enter plant roots from the soil solution by means of complex plant-mediated processes. A continual supply of nutrients at the root surface is required to maintain this uptake. As nutrients become depleted near the active root, supplies of these elements are replenished, principally by three mechanisms that act simultaneously. First, nutrients can move by simple diffusion from areas of high concentration (e.g. near mineral surfaces) to areas of low concentration near the root (fig 25). Second, nutrients may be carried to the root surface in a process known as mass flow. As the root absorbs water, nutrients dissolved in this water are also carried towards the root, in a manner similar to sediment carried by a river. And third, a root may increase the nutrient concentration near its surface by growing into undepleted, nutrient-rich soil. Efficient nutrient uptake by plant roots depends not only on the availability of nutrients but also on factors such as adequate moisture and ease of root penetration. Therefore, a plant may have difficulty obtaining nutrients from a very dry or compacted soil, even if available nutrients are present in abundance.

Soil pH

Hydrogen atoms that have lost an electron are called protons, and it is these species that give a soil its acidic character. The pH of a soil is a measure of this acidity. The pH scale, which ranges from 0 (most acidic) to 14 (least

organic matter are also considered part of the labile pool. The exchangeable nutrients within the labile pool are considered to be plant-available because they can move rapidly into soil solution.

Approximately 90% of a soil's nutrients are unavailable to plants because they are locked up in the soil minerals (e.g. feldspar, mica) and organic matter. These nutrients comprise the stable nutrient pool and can be released only over long periods of time, through mineral weathering or the breakdown of stable organic matter (fig 24). The stable nutrient pool can be considered a long-term reservoir of essential elements.

Unscrupulous or poorly informed vendors at agricultural fairs have been known to peddle powdered materials, composed largely of feldspars and micas, as nutrient-rich fertilizers. Although these materials may contain a high total nutrient content, virtually all of their nutrients are unavailable to plants, acidic), is logarithmic, which means that each unit change in pH represents a ten-fold change in proton concentration (fig 26). For example, a soil of pH 6 will have ten times as much acidity as a soil of pH 7.

Of all soil chemical variables pH is the most important, influencing properties as diverse as nutrient availability, the functioning of micro-organisms and the fate of many pollutants. Soils with pH less than 5 often contain soluble aluminium in amounts that are toxic to plants. These soils are commonly treated with lime to increase the pH and thus reduce the aluminium toxicity. However, plants that require large amounts of iron, such as azaleas and rhododendrons, often prefer the acidic soil conditions under which iron is most available. If these plants are grown in soils of a higher pH, symptoms of iron deficiency may appear.

Soils with pH greater than 9 usually contain sodium in amounts sufficiently large to be detrimental to soil structure. Moreover, plants grown in these high pH soils may show micronutrient deficiencies. The influence of soil pH is shown dramatically by both species and cultivars of *Hydrangea macrophylla*, whose flower colour varies with soil pH (plates 17a, b). Generally, a soil pH between 6 and 7 provides conditions that are the best overall for optimizing nutrient availability (fig 27).

The pH of many soils has been changed as a result of human activity. For example, the application of ammonium-containing fertilizers has reduced the pH of the surface horizons of many soils. Also, nitrogen and sulphur oxides released from vehicles and industry may combine with water in the atmosphere to form strong acids that fall to Earth as acid rain. The ability of a soil to

FIG 26
Soil pH may vary widely. The pH values of common substances are given for comparison.

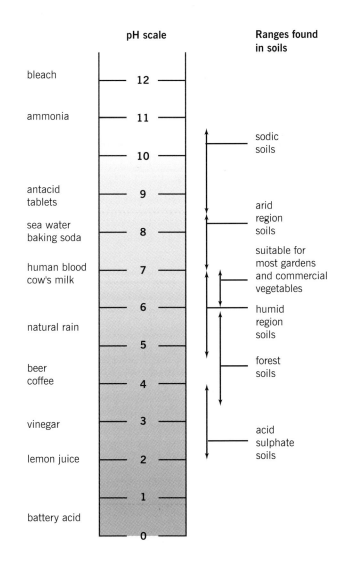

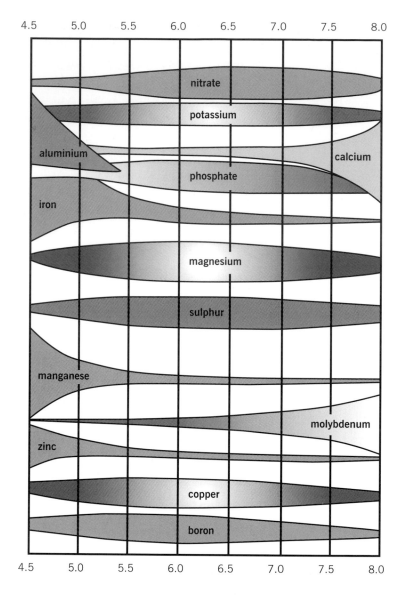

FIG 27
The availability of soil nutrients varies with pH.

CEC to be a soil's nutrient-holding capacity. The CEC is due mainly to the clay minerals and organic matter. The organic matter is especially important because it contributes, on a weight basis, approximately four times as much CEC as the clay fraction. Generally, clayey soils high in organic matter have the highest CEC and fertility. Most garden soils also have high CECs, owing largely to their high organic matter content. Soils containing a large amount of iron and aluminium oxides, such as the Oxisols, have a CEC that is highly pH-dependent, i.e. the CEC of these soils varies with pH, and is highest at high pH.

Nutrient deficiency and toxicity

If all nutrients, except one, are present in amounts sufficient to ensure optimal plant growth, a plant's development can be no greater than is allowed by that one limiting nutrient (plate 17c). A nutrient that limits plant production in this way is therefore called the limiting nutrient, or limiting factor. However, not all limiting factors are nutrients. A plant's growth may be limited not by nutrients, but rather by insufficient water, sunlight or heat. Increasing a factor that is not limiting will do little to increase plant growth. For example, if the growth of a rose bush is limited by insufficient boron, adding nitrogen, water or any other non-limiting factor will do little to improve the plant's health.

As the concentration of a limiting nutrient increases, plant growth will increase correspondingly (fig 28). Once the nutrient has reached a concentration that is optimal for plant growth, the sufficiency range has been reached, and further additions of this nutrient do not increase plant production. If

resist these changes in pH is termed its buffering capacity, and this capacity increases with a soil's lime content and cation exchange capacity.

Cation exchange capacity

Cation exchange capacity (CEC) is the sum total of exchangeable cations that may be adsorbed by a soil. One may consider the

FIG 28

Plant growth is optimized when the concentration of available soil nutrient is within the sufficiency range.

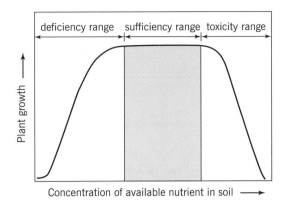

available nutrient concentrations increase beyond the sufficiency range, the plant may take up excess nutrient and symptoms of toxicity will appear. Toxicity leads to reduced growth and even plant death in extreme cases. Nutrient toxicity is almost always confined to the micronutrients. Macronutrient toxicity rarely occurs, even at very high concentrations.

NITROGEN

Nitrogen is a key component of many important biomolecules. The most important of these are proteins, the building blocks of organic tissues; chlorophyll, the molecule essential for photosynthesis; and nucleic acids, which contain the genetic heritage of all organisms. When present in adequate amounts, nitrogen stimulates the growth of roots and above-ground plant tissues, and gives foliage a deep green colour. Despite the importance of nitrogen, this macronutrient is the one most often limiting plant growth. Nitrogen-deficient plants are generally smaller, with spindly stems, and their leaves commonly appear pale green or yellow in a condition known as chlorosis (plates 18a, c).

Nitrogen plays an important role, not only in plant production but also in the health of many ecosystems. For example, nitrogen may occur in soils as nitrate (NO_3^-), a highly mobile species that can be leached into groundwater. Nitrate poses a threat to human health if it is present at high concentrations in drinking water. Also, nitrogen may be released from soils as nitrous oxide gas (N_2O), which contributes to ozone depletion. In view of the importance of soil nitrogen to both plant production and ecosystem health, this element warrants further attention.

The nitrogen cycle

Mineralization – Most soil nitrogen is unavailable to plants because it occurs within slowly decomposing organic matter (fig 29). As micro-organisms attack this organic material, nitrogen is released to solution as inorganic ammonium (NH_4^+) in a process known as mineralization. The ammonium may subsequently become bound to charged mineral surfaces, acquired by plants, or converted to ammonia gas (NH_3), which is then lost to the atmosphere. Only 1 to 4% of the total organic nitrogen in soils of temperate regions is mineralized annually, but this is adequate to support the growth of native vegetation. During the reverse process, known as immobilization, micro-organisms incorporate NH_4^+ into their own tissues, thereby reducing the amount of plant-available nitrogen.

Nitrification – Nitrification refers to the transformation of ammonium to nitrate by bacteria. This transformation occurs via an intermediate, nitrite (NO_2^-), a short-lived

FIG 29
The nitrogen cycle, showing major transformations and pools of nitrogen.

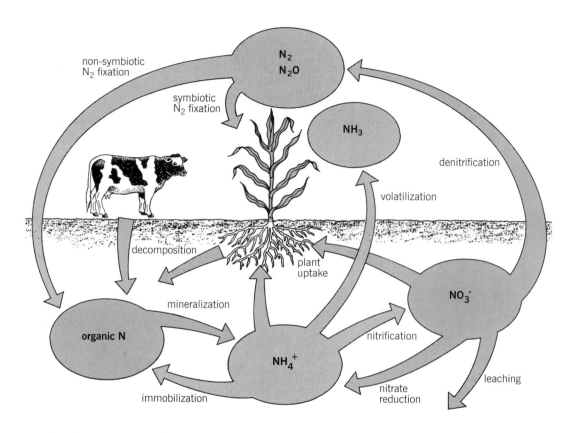

species that is toxic to plants and animals. The nitrate thus produced may be acquired by plants or lost from the soil by leaching. Nitrate leaching is troublesome not only because this valuable nutrient is lost but also because of the potential for groundwater contamination. The problems associated with leaching of excess nitrate have prompted the development of various synthetic compounds that inhibit nitrification. These substances reduce the activity of the nitrifying bacteria, thereby maintaining a greater proportion of plant-available nitrogen as ammonium, a much less mobile form.

Denitrification – Nitrogen may be lost from soil not only through nitrate leaching but also by conversion of nitrate to various gaseous

forms, such as N_2O and N_2, in a process known as denitrification. Under conditions of poor aeration, certain bacteria transform nitrate into these gaseous forms, which are then released to the atmosphere. Denitrification losses are therefore greatest in rice paddies, natural wetlands and other flooded soils (plate 18b). Even in relatively well-drained soils, there may be anaerobic zones within some aggregates where conditions are favourable for denitrification. Losses of nitrogen by denitrification can be rapid and significant. Estimates indicate that 10 to 20% of all nitrate originating from fertilizer may be lost by denitrification.

Nitrogen fixation – Dinitrogen gas (N_2) in the atmosphere may be captured by soil micro-organisms and converted into forms that can be used by plants. These microbes may fix N_2 either symbiotically (i.e. in association with plants) or non-symbiotically. Most N_2 fixation occurs symbiotically. Interestingly, the symbiotic legume–*Rhizobium* associations may fix enough N_2 to meet nearly 80% of the legume's nitrogen needs. Considerable soil nitrogen additions can therefore be achieved by returning to the soil the entire legume plant while it is still green, in a practice known as green manuring.

Any given nitrogen atom resides in an animal or plant for only a short time, as it is moved through the nitrogen cycle by a myriad of chemical and biological reactions. It is thought-provoking to realize that the nitrogen within the tissues of each of us was at various times a component of the atmosphere, soil organisms and humus.

PHOSPHORUS

Phosphorus is essential for the normal functioning of cell membranes, and it is a key component of both nucleic acids and adenosine triphosphate (ATP), the molecule that fuels nearly all metabolic processes. Phosphorus deficiency in plants is not always easy to recognize, although it may result in reduced flowering, inferior seed quality and delayed plant maturity (plates 19a, b).

Unlike nitrogen, soil phosphorus does not occur in gaseous forms or forms that are toxic. Like nitrogen, however, phosphorus is important in both ecosystem health and plant production. For example, insufficient plant-available phosphorus will lead to reduced vegetative cover and increased risk of soil erosion. Conversely, too much available phosphorus can lead to degradation of lakes and rivers through a process known as eutrophication (see box). Archaeologists are also interested in soil phosphorus. Enrichments of this element in a profile may indicate the presence of phosphorus-rich skeletal remains of humans or other animals.

Sources of soil phosphorus

In contrast to soil nitrogen, which is found almost exclusively within the organic fraction, soil phosphorus is distributed approximately equally among the organic and inorganic fractions. Organic soil phosphorus occurs within a diverse group of organic molecules that have been synthesized by micro-organisms, or that originate from plant and animal residues. Phosphorus held in the organic fraction is made available to plants through mineralization, in a manner similar to that which increases nitrogen availability.

In soils of high pH, inorganic phosphorus occurs mainly as apatite, a very insoluble calcium phosphate mineral. Apatite solubility increases rapidly as pH decreases. Consequently, this mineral may be absent from highly weathered acidic soils. In these acidic soils, phosphorus forms highly insoluble minerals with iron and aluminium.

Phosphorus availability and uptake

Phosphorus must be present in solution as phosphate anions ($H_2PO_4^-$ and HPO_4^{2-}) or, rarely, soluble organic complexes before it can be absorbed by plant roots. Because of the very low solubility of both organic and inorganic phosphorus sources, deficiencies of these plant-available forms are common, even in the presence of large total amounts of soil phosphorus.

TOO MUCH OF A GOOD THING – EUTROPHICATION

Run-off waters and eroded sediments from animal feedlots and phosphorus-amended soils often contain large amounts of phosphorus. This water and sediment frequently collect in rivers and streams where phosphorus is the nutrient limiting plant growth. The addition of phosphorus to these phosphorus-limited systems is known as eutrophication, and this causes a rapid growth of algae and other plants to produce algal blooms (plate 19c). These blooms are problematic because certain algal species, such as *Pfiesteria*, produce deadly toxins. Also, when the abundant biomass eventually decomposes, much of the oxygen dissolved in the water is consumed. The resulting anoxic conditions kill many fish and other aquatic organisms, thereby reducing ecosystem health. Eutrophication can be reduced by adopting management practices that control soil erosion and therefore minimize the loss of nutrient-rich waters and sediments from soils.

Efforts to increase plant-available phosphorus by applying manures or inorganic fertilizers are not always successful. The phosphorus released from these relatively soluble sources is quickly converted to insoluble forms or fixed tightly to iron and aluminium oxides. Often only 10% of the phosphorus in these amendments is available to plants during the season of application. A practice commonly employed among more affluent farmers is to add each year more phosphorus than can be removed by the crop. Once the soil's capacity to fix phosphorus has been saturated, plant-available phosphorus levels may then increase significantly.

Management of phosphorus on the highly weathered soils of many developing countries presents special problems. These soils, with their large iron and aluminium oxide content, have a very high phosphorus-fixing capacity. Moreover, farmers in these regions rarely have the resources to apply large amounts of phosphorus, as do their more affluent counterparts. Consequently, in these highly weathered soils, the organic fraction may be the most important source of plant-available phosphorus because most of the inorganic sources have been weathered from the profile. Careful management of the organic fraction is therefore of utmost importance to maintain adequate amounts of plant-available phosphorus.

Phosphorus availability can be increased in all soils by mycorrhizae. These symbiotic associations between roots of many plants and certain fungi greatly increase phosphorus uptake by: (i) increasing root surface area, (ii) releasing organic acids that increase the solubility of phosphorus-containing minerals, and (iii) producing enzymes that facilitate the mineralization of phosphorus-rich organic matter.

POTASSIUM

After nitrogen and phosphorus, potassium is the nutrient most likely to limit plant growth. Potassium is an essential plant nutrient because it is, among other things, a key component of many important enzymes. It is also essential for optimal water use efficiency, conferring to the plant considerable drought tolerance. This nutrient also enhances the colour of flowers, and improves the taste and texture of many fruits and vegetables. Bananas and potatoes have an especially high requirement for potassium.

Deficiencies of potassium are generally easier to recognize than those of phosphorus. Insufficient potassium causes the leaf margins to appear chlorotic initially, and then finally die, in what is termed necrosis (plates 20a – d). These symptoms are most visible in the older leaves, as potassium is shunted to the younger tissues of the plant. An excess of soil potassium is of limited concern because this element is not toxic and does not cause environmental problems such as eutrophication.

Sources of soil potassium

Soil potassium originates almost entirely from the inorganic components, principally micas and feldspars. As these minerals weather, potassium is released slowly from their framework structures into forms that are plant-available. There is often enough potassium released from these minerals to satisfy the requirements of crops for decades or centuries. For example, most of the Mollisols that occur on the Canadian prairies have developed from potassium-rich parent materials. Consequently, these moderately weathered soils have remained productive for nearly a century, despite the lack of potassium fertilization. Conversely, highly weathered soils, such as Ultisols and Oxisols, generally require potassium amendments because the micas and feldspars have largely been lost from the profile.

Potassium occurs also in soil organic matter. However, unlike nitrogen and phosphorus, potassium is not a major structural component of organic tissues. Rather, most potassium exists within cells as a soluble cation (K^+) that can be leached rapidly from the tissues even in the absence of microbial breakdown. Therefore, only fresh organic residues contain appreciable amounts of potassium.

Potassium availability and fixation

Potassium must be present in the soil solution as the positively charged cation (K^+) before it can be taken up by plant roots. As K^+ is removed by plants from the soil solution, it is replenished by exchangeable K^+ moving from the surfaces of minerals and organic matter. The K^+ in solution, together with the exchangeable K^+, comprise the 'readily available' potassium. Although the readily available potassium is less than 2% of the soil's total potassium content, it is this small but available nutrient pool upon which a plant's survival depends. Soils with very low cation exchange capacities, such as sandy soils with little organic matter, may have difficulty maintaining the solution K^+ concentration at adequate levels.

Layer silicates such as vermiculite and illite may bind K^+ within their layered structures with such strength that this bound potassium is no longer readily available to plants. This phenomenon is known as potassium fixation and is promoted by repeated cycles of wetting/drying and freezing/thawing. Although K^+ fixed in this manner can no longer move readily into the soil solution, it is an extremely important source of slowly available potassium. Approximately 95% of all soil potassium is unavailable to plants. This potassium is locked in the soil minerals, principally the micas and feldspars.

Potassium cycling

If the upper part of a soil becomes potassium deficient, deep-rooted plants (e.g. trees and many grasses) can be used as nutrient pumps to translocate potassium from the deeper levels, through the plant tissues, to the soil surface. Leaves and other plant debris that collect at the surface are quickly leached of their potassium, which then enters the soil. The cycling of potassium in a soil is governed primarily by this nutrient pumping, as well as the soil's cation exchange capacity and mineral weathering rates.

SULPHUR

Sulphur is an essential component of many amino acids and vitamins. Many of the compounds responsible for the distinctive taste and smell of onions and some cabbages also contain sulphur. These plants, as well as corn, cotton and sorghum, have high sulphur contents and, consequently, high sulphur requirements.

Sulphur deficiency results in spindly, slow-growing plants with chlorosis that appears first in the youngest leaves (plates 21a, b). Deficiencies of sulphur are most common in sandy soils with little organic matter, highly weathered soils of the tropics, or where sulphur additions from atmospheric sources are small.

Sources of sulphur

There are three principal sources of soil sulphur: organic matter, minerals, and atmospheric gases and particles. In most soils the majority of sulphur occurs within the organic fraction. Mineralization of this organic material by micro-organisms releases to solution the sulphate anion (SO_4^{2-}), which is easily absorbed by plant roots. Sulphur also occurs in soil minerals, principally the sulphides and sulphates. The sulphates (e.g. gypsum: $CaSO_4$) are often abundant in soils of arid regions. In the lower horizons of these soils, where organic matter is scarce, the sulphates may provide most of the plant-available sulphur.

Significant amounts of sulphur may also be added to a soil through atmospheric deposition. Sulphur is released to the atmosphere, in varying forms and amounts, from volcanic eruptions, industrial emissions or the burning of vegetation. This atmospheric sulphur may then fall to Earth as precipitation or dust. Plants acquire atmospheric sulphur either directly, by absorption through foliage, or indirectly through root uptake. Many soils in eastern North America, especially those that are downwind from industrial sources, receive considerable amounts of sulphur from atmospheric deposition. Consequently, few soils in these areas are sulphur deficient.

Sulphur and soil acidity

Although sulphur is an essential nutrient, it is the cause of much acidity in some soils. For example, various sulphur gases in the atmosphere may combine with water to produce sulphuric acid (H_2SO_4), which falls to Earth as acidic rain. Nitrogen gases, it should be noted, may also react with water to produce nitric acid (HNO_3), which also contributes to acid rain. Moreover, certain minerals may contain sulphur in a form that can be oxidized to produce additional acidity. For example, sulphur in pyrite (FeS_2, fool's gold) can be readily oxidized to produce H_2SO_4, giving rise to acid sulphate soils. These soils commonly develop in recently drained coastal areas, where such H_2SO_4 accumulations may lower the soil pH to less than 2.

CALCIUM AND MAGNESIUM

Calcium deficiencies are rare because this element is present in many minerals that dissolve easily to release calcium cations (Ca^{2+}), the form taken up by plants. In many soils calcium is by far the most abundant exchangeable cation. However, calcium deficiencies (plate 21c) may occur in plants grown on very sandy soils, or soils that are acidic and have been extensively leached. Peanuts, which have high calcium requirements, are especially susceptible to calcium deficiencies on sandy or weathered soils.

Magnesium is absorbed by plants as the Mg^{2+} cation, which is required for the normal functioning of chlorophyll, an essential light-absorbing pigment. When this nutrient is deficient in soils (plates 22a – c), Epsom salt ($MgSO_4$) may be applied as a soluble and readily available magnesium source.

MICRONUTRIENTS

The nine micronutrients, also known as trace elements, listed in Table 4 are as essential for plant growth as the macronutrients, but they are required in smaller amounts. The main function of micronutrients, with the exception of chlorine, is one of enzyme activation. Inadequate supplies of these essential elements can therefore lead to a disruption of important metabolic pathways.

The micronutrients cannot be easily translocated within plants from older to newer tissues. Consequently, inadequate micronutrient availability results in deficiency symptoms that are most apparent in the youngest plant tissues, such as new leaves and flowers (plates 23a, c).

Sources and availability of micronutrients

Micronutrients are derived initially from the weathering of soil minerals. Once these elements are released to solution, they are either taken up by plants, adsorbed by organic or mineral surfaces or leached from the soil profile. Those micronutrients that occur in solution as cations (i.e. Fe, Cu, Mn, Zn, Ni, Co) may combine with soluble organic molecules to form specialized associations known as chelates. These chelates maintain the cations in solution, thereby increasing micronutrient availability. The organic molecules required for chelate formation may originate from humus or plant roots, or they may be added as synthetic amendments in a deliberate attempt to increase micronutrient availability.

Organic soils, sandy soils or soils that are highly leached generally have the lowest total micronutrient content. These soils may benefit from the application of micronutrients in the form of inorganic fertilizers or animal manures. In soils with adequate total micronutrient content, pH is the most important variable determining their availability. Under acidic conditions, iron, copper, manganese, zinc,

BORON IN PLANT NUTRITION

Boron is among those micronutrients that are most often deficient. The crops with the greatest sensitivity to boron deficiency are celery, sugar beets, apples, tomatoes and grapes. Boron deficiency reduces the quality of flowers and fruit (plate 23b). For example, apples and tomatoes that are boron deficient develop scaly surfaces, and their internal tissues are often cork-like. Boron-deficient grapes are small and of poor quality. In celery, boron deficiency results in cracked stems.

Sodium tetraborate (borax) is the compound most commonly applied to alleviate boron deficiencies. Various borax formulations are now routinely applied to many economically important crops, such as apples, tomatoes and grapes. The amount of boron applied to the crop must be strictly controlled, because there is a narrow concentration range over which boron (and each of the other micronutrients) is neither toxic nor deficient.

nickel and cobalt are abundant in soil solution. If the pH is sufficiently low, these metals may occur in solution at concentrations that are toxic to some plants. An increase in pH, conversely, decreases the solubility and availability of these elements. Consequently, soils of high pH, such as Aridisols and many other soils of dry regions, may exhibit micronutrient deficiencies. Molybdenum, in contrast to the other micronutrients, becomes more soluble and available as pH increases.

When optimizing micronutrient availability, extremes of pH must be avoided. Generally, a medium-textured soil between pH 6 and 7 will provide all the micronutrients needed for normal plant growth.

SOIL FERTILITY TESTING

"How dangerous it is to reason from insufficient data"

Sherlock Holmes

Reduced plant growth may be caused by, among other factors: soil compaction, poor aeration, insufficient moisture or sunlight, diseases and pests, as well as inadequate soil nutrient status. Visual symptoms of nutrient deficiency, such as chlorosis and poor fruit quality, provide a useful first step in identifying those nutrients that may be lacking. However, a more quantitative determination of nutrient availability is often required, especially if specific recommendations for fertilizer additions are to be made. Various soil fertility tests have therefore been devised to assess the level of plant-available soil nutrients.

Sampling

Soil sampling is the first, and perhaps most important, step in soil analysis. The samples must be representative of the area of interest. If, for example, soil from a vegetable garden is to be sampled, several samples should be obtained and then pooled to reduce the spatial variability that is likely to exist. Also, soil from vegetable gardens, flower beds and lawns should be sampled separately, because each of these three areas will have different nutrient demands.

Because of seasonal variations in nutrient uptake, mineralization and other factors that govern nutrient cycling, tests for nutrient availability may give widely different results depending on the time of sampling. The best time for sampling is generally thought to be as near as possible to planting. The depth of sampling is usually about 20 cm, although some soil nitrogen tests require sampling depths as great as 90 cm.

Testing

Soil tests are chemical analyses that are designed to measure the amount of plant-available nutrient. This is a considerable challenge. As we have seen, plants incorporate various strategies, such as mycorrhizae, to increase nutrient uptake. A good soil test will mimic these biological strategies and measure the plant-available nutrient quickly and reliably.

A soil test typically involves mixing a small amount of soil with a dilute acid, salt solution or chelating agent for a fixed time, then measuring the nutrients that have been dissolved in the extracting solution (fig 30). Soil tests for phosphorus, potassium, calcium and magnesium have been developed and they are generally reliable for a range of soil types and climatic conditions. Tests for available nitrogen and sulphur have also been developed but, because of the many biological factors governing the transformations of these nutrients, the reliability of these tests is not always satisfactory. Plant-available micronutrients are commonly extracted with various chelating agents.

Local horticultural societies can usually recommend reliable soil testing labs. Such laboratories will, when making fertilizer recommendations, take into account current nutrient levels, the expected nutrient demand, and other factors such as soil texture and pH.

SOIL AMENDMENTS

The growth of human populations has placed increasing demands on the nutrient resources of soils. For example, the production and harvest of high-yielding crops needed to feed this growing population has resulted in the removal of large amounts of nutrient, much of which is not returned to the soil. Fruits and vegetables grown in one's garden may

FIG 30
Once the soil and the extracting solution have been mixed for a suitable time, the slurry is filtered to obtain the solution and dissolved nutrients, which are then measured by various laboratory techniques.

similarly remove from the soil more nutrient than is returned. Likewise, grass clippings and other garden wastes contain nutrients that commonly end up in rubbish bins and landfill sites. Such nutrient losses from soil cannot be sustained indefinitely.

The soil tests described previously may recommend the addition of certain nutrients in order to optimize plant growth. These nutrients may be obtained from external sources, such as organic and inorganic fertilizers, or simply by improving soil management practices. This latter method of nutrient enrichment is often the most desirable and cost-effective. One may, for example, improve a soil's fertility by increasing the amount of crop residue that is returned to the soil. This practice is especially effective with legumes, such as peas, clover and alfalfa, which may fix considerable amounts of atmospheric nitrogen.

Changing the type of crop that is grown each year on a particular field can improve soil fertility. The nutritive benefits of these crop rotations may be caused by the different residue types, nutrient requirements and rooting patterns of each plant type. Crop rotations are beneficial also because they interrupt the cycle of diseases, weeds and other pests. However, management practices alone are often unable to provide nutrients in amounts sufficient to meet the needs of high-yielding crops. In these cases, nutrients from external sources must be added.

Inorganic amendments

Much of the increase in worldwide food production during the last century can be attributed to increased inorganic fertilizer use. The nutrients most commonly applied to soils are nitrogen, phosphorus and potassium.

Nitrogen – Inorganic nitrogen fertilizers are produced by means of an energetically expensive procedure that captures atmospheric nitrogen to produce ammonia gas. This gas may be pressurized to form a liquid, anhydrous ammonia, which is then applied to soils. Many other nitrogen fertilizers, both liquids and solids, can be derived from this ammonia. All of these inorganic fertilizers release nitrogen in forms that are readily available to plants. The most common nitrogen fertilizers are listed in Table 5.

Phosphorus and potassium – Phosphorus and potassium fertilizers are derived from ancient

TABLE 5
Nutrient content of several common soil amendments

	Percentage by weight					
Inorganic amendments	N	P	K	S	Ca	Mg
Anhydrous ammonia [82–0–0]	82					
Urea [46–0–0]	46					
Ammonium sulphate [21–0–0(24)]	21			24		
Diammonium phosphate [18–46–0]	18	20				
Potassium chloride [0–0–60]			50			
Calcitic limestone					36	
Organic amendments*						
Poultry manure	7	2	2	1	2	2
Swine manure	3	1	1	1	2	1
Cattle manure	4	1	3	1	1	1
Sewage sludge	5	2	1	1	2	1
Compost	2	1	1	1	3	1

*(average of dry weight; composition variable)

mineral deposits. Phosphorus is mined as the mineral apatite, which is treated with various acids to produce fertilizers containing soluble forms of phosphorus. Potassium is extracted from salt beds, principally as potassium chloride (KCl), which is then purified before use as a fertilizer. The world's largest known reserves of potassium chloride are currently being mined in Saskatchewan, Canada. Diammonium phosphate and potassium chloride are among the most commonly applied phosphorus and potassium fertilizers (Table 5).

Liming – Soils with low pH have many undesirable characteristics. At pH less than 5, soils may suffer from toxicity of aluminium, manganese and occasionally iron. These acidic soils commonly show deficiencies of calcium, magnesium and molybdenum. The activity of micro-organisms is also reduced in acidic soils, resulting in a slower rate of nitrogen and phosphorus mineralization. Moreover, the decreased rate of microbial activity at low pH adversely affects soil structure because the organic materials needed for the formation of stable aggregates are not produced in sufficient amounts.

Soil pH can be increased with the application of lime, a material composed of various acid-neutralizing compounds. Most agricultural lime is composed of crushed limestone, calcium carbonate ($CaCO_3$). The more finely crushed the limestone, the more quickly it will react with the soil to increase pH. The quantity of lime required depends on the type of lime used, the desired increase in pH and the buffering capacity of the soil. Buffering capacity increases with the amount of organic matter and clay. Losses

of calcium by plant uptake and leaching, particularly in humid regions, necessitates repeated applications of lime to maintain the desired pH.

Fertilizer grade – The nutrient content of a fertilizer is indicated by three numbers, which comprise the fertilizer grade (fig 31). For example, a fertilizer may have a grade of 16–8–24. These numbers indicate that this fertilizer contains, on a weight basis, 16% elemental nitrogen, 8% phosphorus (expressed as available P_2O_5), and 24% potassium (expressed as soluble K_2O). The expression of phosphorus and potassium content as P_2O_5 and K_2O is an unfortunate convention that is slow to change. One reason for the reluctance to express phosphorus and potassium on an element percentage is that the fertilizer would then appear to be of a lower grade. For example, by expressing all nutrients on an element percentage, this 16–8–24 fertilizer would become 16–3–20.

Fertilizers that contain significant amounts of sulphur are assigned a four-digit grade. The fourth number, often in parentheses, indicates the percentage elemental sulphur. A fertilizer

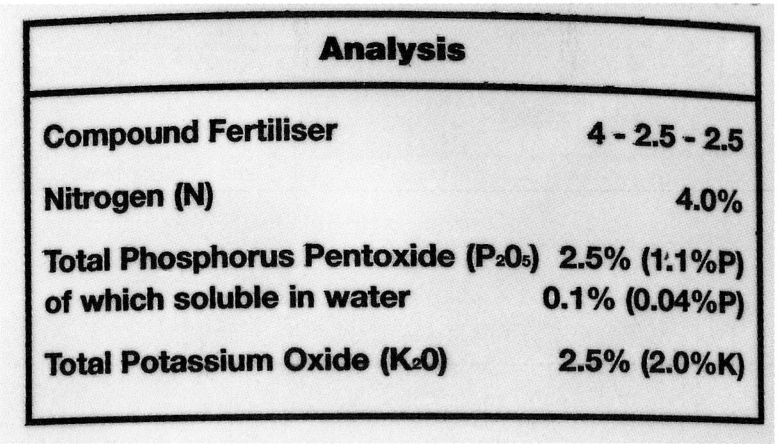

FIG 31
A typical fertilizer label, showing the fertilizer grade and the element percentages of phosphorus and potassium.

Analysis	
Compound Fertiliser	**4 - 2.5 - 2.5**
Nitrogen (N)	**4.0%**
Total Phosphorus Pentoxide (P_2O_5)	**2.5% (1.1%P)**
of which soluble in water	**0.1% (0.04%P)**
Total Potassium Oxide (K_2O)	**2.5% (2.0%K)**

COMPOSTING – FROM RAGS TO RICHES

Composting refers to the biologically mediated decomposition of organic materials to produce humus-like substances outside of the soil. This process is similar to that which occurs within the soil, except that composting is much more rapid and may generate considerable heat. Composting has several benefits: (i) nutrient elements are concentrated, (ii) most weeds and pathogenic organisms are destroyed, (iii) many toxic organic compounds are degraded, and (iv) a stable organic material is produced. The final product, compost, can be used as a potting media, slow-release fertilizer and soil conditioner.

Compost can be prepared from many materials: grass clippings, leaves, kitchen waste and sawdust, as well as plant residues from vegetable gardens and flower beds. The carbon to nitrogen (C/N) ratios of these different materials vary considerably. For example, the C/N ratio of sawdust may be 400:1, whereas that for the

most nitrogen-rich wastes (e.g. manures) may be as low as 10:1. It is desirable to have a starting material with an overall C/N ratio of less than 30:1. To achieve this ratio, one may need to incorporate nitrogen-rich wastes, or add a small amount of nitrogen fertilizer.

During the composting process, CO_2 is released as micro-organisms degrade the organic material (fig 32). To ensure that enough of these decomposer organisms are present initially, it is advisable to add a small amount of garden soil to the original compost heap. The loss of carbon during composting gives rise to a finished product with a C/N ratio near 15:1.

The compost pile must be moist and well aerated at all times (plate 24a). A pile is at the correct moisture level when it is damp, but not dripping wet. Occasional turning of the compost heap will ensure sufficient aeration to maximize microbial activity. The intense metabolic activity during successful composting may elevate temperatures in the centre of the pile to as high as 75°C. Such temperatures facilitate the destruction of weeds and pathogens. The time required to prepare a stable compost ranges from several weeks to several months, depending mainly on moisture levels, oxygen availability, pH and C/N ratios of the starting materials.

FIG 32
Composting involves several inputs and outputs.

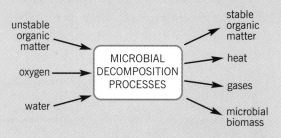

with the grade 16–20–0(14) would therefore contain 14% elemental S, 16% elemental N and 20% P_2O_5.

Organic amendments

Only relatively recently have inorganic fertilizers become primary sources of plant nutrients. Throughout most of agricultural history, organic amendments have been the main external source of plant nutrients for crop production. For example, ancient writings indicate that manuring was a common agricultural practice in Greece nearly three thousand years ago. Even today, organic amendments remain an essential source of nutrients for crop production in resource-poor developing countries. Additionally, gardeners from many parts of the world have experienced the horticultural benefits that can arise from the application of well-prepared compost (see box).

Animal manures – Both the solid and liquid fractions of animal manure are important sources of nutrients (plate 24b). However, nearly all of the phosphorus occurs within the solid fraction, which also provides valuable organic material. The nutrient composition of manures varies widely depending on, among other factors, the type of animal, the quality of the feed and how the manure is stored prior to application. Most manures provide, in addition to nitrogen, phosphorus and potassium, appreciable amounts of micronutrients. However, the large quantities of soluble salts in some manures demand the prudent use of these amendments when applied to soils vulnerable to salinization.

Sewage sludge – The solid material removed during wastewater treatment is known as sewage sludge, or biosolids. Sewage sludge, like manure, is an important source of plant nutrients. The nutrient content of sludge is comparable to that of manures, although sludges generally contain more micro-nutrients. Most sludge is applied to land as a slurry, or sometimes as a partially dried solid, which minimizes transport costs. A significant challenge in utilizing sludge produced by large cities is finding sufficiently large areas of nearby land suitable to receive this amendment. Another important consideration is to ensure that the sludge does

not contain heavy metals in amounts that could accumulate in soils to levels that are toxic to humans.

Organic versus inorganic amendments

Inorganic fertilizers have several advantages over organic fertilizers. First, inorganic amendments provide, on a weight basis, much more nutrient than organic amendments. For example, 100 kg of a 10–5–10 grade fertilizer provides as much nitrogen, phosphorus and potassium as 2 tonnes of an average farm manure. Also, because much of the weight of this manure is water, transport costs are considerably higher for the organic amendment. Moreover, the highly variable composition of organic fertilizers makes calculating application rates less precise. Second, nutrients within inorganic fertilizers are generally much more readily available than those from organic sources. And third, manures and sludges may, if not properly treated, contain many weeds and pathogenic organisms.

However, organic nutrient sources have several important benefits not provided by inorganic amendments. Organic materials, especially those that are highly decomposed, improve soil structure, water-holding potential and cation exchange capacity. These benefits are particularly valuable where the surface horizon has been lost or degraded, as through erosion. Also, because most organic amendments are free of cost, they may provide the only economically viable source of nutrients for small-scale farmers in developing countries and elsewhere.

The high cost of inorganic fertilizers and other agricultural inputs has prompted many small-scale farmers to adopt a system of nutrient management known as low-input sustainable agriculture. Such a system attempts to produce crops sustainably with the minimal use of inorganic fertilizers. Farmers that adopt this method of production recognize the importance of inorganic nutrient sources, yet their efforts to increase crop yields focus mainly on improving crop rotations, incorporating better crop varieties and increasing the amount of nutrient from organic fertilizers. Despite such comprehensive management strategies, yields from these low-input farms tend to be only 60 to 90 percent that of conventional farms. However, the lower cost of inputs may result in a net increase in profits for these small-scale farmers.

Most nutrients added to a soil, whether from organic or inorganic sources, are not taken up by the plant during the season of application. Rather, the added nutrients become incorporated into the complex nutrient cycles described previously. The most immediate benefit of fertilizer addition is to stimulate biological activity and nutrient cycling, and in this way increase the availability of existing nutrient pools. Once nutrients are in solution in available form, those derived from organic sources are indistinguishable from those derived from inorganic sources.

SOIL USE AND MISUSE

Soil mismanagement and degradation have contributed to the decline of once prosperous societies. The civilizations that flourished in ancient Greece and Rome relied heavily on the fertile, but fragile, soils that occurred on the hilly lands of the region. The Romans, despite their well-documented reverence for *mater terra*, frequently adopted a more utilitarian and short-sighted approach to land management. Consequently, soils on steep slopes became heavily eroded and lost much of their productivity. Unfortunately, soil degradation is a problem of the present as well as the past.

QUALITY AND DEGRADATION

In recent decades, soil quality and degradation have been the focus of much attention among both scientists and the general public. Assessing soil quality, and identifying the factors responsible for soil degradation, are prerequisite to effective soil management. The discussion will now turn to these two very important concepts.

Quality

The increasing use of the term soil quality by scientists, farmers, environmentalists and politicians, reflects a growing awareness of the importance of soils in food production and ecosystem health. Soil quality, also known as soil health, refers to the capacity of a soil to promote the health of plants, animals and humans, maintain environmental quality, and sustain biological productivity. Thus, soil quality refers to a capacity to perform certain health, environmental and productivity functions.

Assessing soil quality can be difficult because each soil function requires a different set of soil properties. Despite this difficulty, there is currently much effort to develop specific and reliable soil quality indices. These indices would provide a quantitative measure of a soil's capacity to perform certain functions, such as supporting plant growth or remediating pollution.

Soil quality assessment can be considered analogous to a medical examination by a doctor. The doctor will measure variables such as blood pressure, temperature and lung capacity before diagnosing overall health. Likewise, the soil scientist may measure organic matter content, cation exchange capacity or water-holding potential before assessing soil quality. Rarely, if ever, can an assessment of either human health or soil quality be made on the basis of one test alone. A variety of soil parameters must first be obtained, and then interpreted in light of the intended use of the soil.

Degradation

Soil degradation occurs when a soil is altered in such a way that its capacity to perform a certain function is diminished. Degradation leads to reduced soil quality. Because of the complexity of soils, degradation may occur in many ways and be caused by many factors. It is convenient, however, to assign degradative

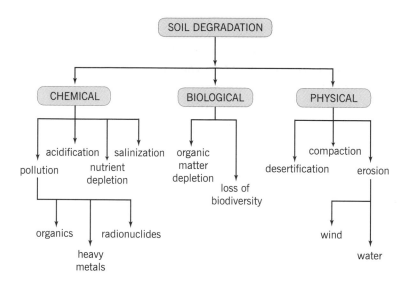

FIG 33
Soil degradation takes
many forms.

processes to one of three categories: physical, biological or chemical (fig 33). Physical degradation includes compaction, desertification, and erosion by wind or water. Much physical degradation of soils is associated with desertification, a process whereby spreading desert conditions reduce the quality of arid and semiarid soils (plate 24c). Biological degradation encompasses loss of organic matter, and reduction of soil organism abundance and diversity. Chemical degradation may occur through salinization, acidification, nutrient depletion, or the addition of organic or inorganic contaminants.

Many of these degradative processes are closely related. For example, desertification is associated with a loss of organic matter and vegetation, which in turn increases nutrient depletion and susceptibility to erosion. The following sections will examine the role that erosion, salinization and anthropogenic pollutants play in soil degradation.

SOIL EROSION

Erosion of land surfaces is a natural process that has been occurring for millennia. The rate of this erosion is, however, generally very slow in the absence of human activity. For example, the erosion rate of most vegetated, undisturbed soils is slower than the rate of their formation, which is about 1 cm every 100 to 400 years. The mere presence of soils is proof that they form faster than they erode.

Disruption of vegetation or the soil surface by humans or animals can greatly increase the rate of erosion. This accelerated erosion may be more than 100 times as rapid as that which occurs naturally. The cutting of forests on hillsides, especially in areas of high rainfall, may cause rapid loss of soil through water erosion. Similarly, overgrazing or excessive tillage in arid and semiarid regions can greatly increase soil losses through wind erosion. The devastation caused to many soils on the North American plains during the 1930s is a testament to the destructive forces of wind.

The effects of soil erosion may be confined to the soil itself, or they may appear off-site, far from the place of erosion. Regardless of where the damage occurs, the far-reaching consequences of soil erosion result in significant social, economic and environmental costs.

On-site losses

The soil surface horizon, which is rich in organic matter and nutrients, is constantly exposed to the erosive forces of wind and water. Loss of soil through erosion reduces soil quality significantly. Erosion is destructive not only because it may remove vast amounts of soil but also because it preferentially removes the most valuable components: organic material and the fine mineral particles. These losses reduce cation exchange and water-holding capacities, and also decrease biological activity.

In cases of severe erosion, the entire surface horizon may be removed, thereby exposing the underlying B or C horizon. In such extreme instances, the concomitant deterioration of soil structure greatly reduces water infiltration rate. Consequently, surface run-off is increased and erosion is accelerated further.

Distant effects of soil erosion

Many deleterious effects of soil erosion appear downstream or downwind from the site of erosion. Sediments eroded from soils may enter streams and lakes where they increase turbidity and disrupt aquatic ecosystems. These sediments may also collect in reservoirs, resulting in reduced water storage capacity. The cost of recovering these reservoirs through dredging is often many times higher than the cost of implementing measures that would control soil erosion and prevent this costly sediment accumulation. Sediments from eroded soils also frequently contain much phosphorus, which contributes to eutrophication, and also pesticides, which may contaminate water used for recreation or drinking.

Wind erosion of soils may also have far-reaching consequences. The clay-sized particles within the dust clouds present a human health hazard, as they irritate the lining of lungs, causing inflammation and other damage. Moreover, the dust storms resulting from soil erosion during dry, windy weather may create hazardous driving conditions on motorways, often with tragic consequences.

In light of the many direct and indirect consequences of soil erosion, protecting this fragile resource must be a priority. Fortunately, the environmental, social and economic costs of erosion can be controlled by implementing sound soil management practices. The effective implementation of these practices requires that one first have an understanding of the causes and mechanisms of soil erosion.

Water erosion

Large raindrops fall to the Earth's surface at about 30 km/h. Despite the small mass of a single raindrop, this high velocity gives the drop considerable kinetic energy, which is dissipated upon impact with the soil surface (fig 34). Such violent collisions detach soil particles, particularly fine sands and silts, displacing them up to 2 m from the point of impact. Sustained attack by raindrops may destroy aggregates, especially those that have been weakened by prolonged wetting. When the disaggregated material dries, a hard crust may form at the soil surface. These crusts often inhibit seedling emergence and restrict water infiltration, thereby encouraging run-off during subsequent rainfall.

If a rainstorm is of sufficient duration and intensity, the soil's infiltration capacity will be exceeded and water will begin to flow over the surface. The particles detached during raindrop impact will then be transported downslope by the flowing water. If the water flows smoothly, then the detached particles will be removed from the soil uniformly. This type of erosion is known as sheet erosion

FIG 34
Raindrops fall to the Earth's surface with considerable force, disrupting soil aggregates and dispersing individual particles.

(plates 24d, 25a). If, however, the water begins to flow preferentially in lower places and small depressions, erosion will occur by channelized flow. Water flowing in this manner moves faster and becomes turbulent, thus scouring the soil surface and accelerating erosion. Channelized flow occurs initially in small channels called rills. As this rill erosion proceeds (plate 25b), the channels enlarge until they form gullies, which allow for increasingly faster and more turbulent flow, leading ultimately to gully erosion (plate 25c).

Rills and gullies differ principally with respect to size. Rills can be easily removed by tillage, whereas gullies are much larger, often preventing the use of tractors. Although gullies are the most visible form of water erosion and often scar the landscape severely, most soil is moved by sheet and rill erosion.

Control of water erosion – Water erosion can be reduced by controlling soil detachment and transport. Detachment is controlled most effectively by maintaining vegetative cover. Undisturbed grasslands and forests provide the greatest protection. For soils that have been disturbed, such as intensively farmed cropland, conservation tillage systems can be implemented to maintain a layer of crop residue (plate 26a). This surface mulch of organic material serves to disperse the energy of raindrop impact.

Soil transport, which is most problematic in hilly areas, is controlled by reducing the flow of water downslope. Terraces may be constructed to reduce slope gradient and thus minimize water flow and soil transport (plate 26b). However, because of the high cost of construction and maintenance, terracing is recommended only for intensively farmed areas where arable land is in short supply. Contour cultivation, which involves tilling and planting at right angles to the natural slopes of the field, can effectively reduce soil transport on moderately hilly land.

Wind erosion

Wind erosion, like water erosion, is a global problem. Every continent has soils that have been degraded by wind erosion. Although wind erosion occurs principally in arid and semiarid regions, soils of humid climates that experience occasional drying are also vulnerable to the erosive forces of wind.

Wind erosion becomes significant only when ground-level wind velocity exceeds 25 km/h. As air moves rapidly over the soil surface, small grains become detached from aggregates and clods. Although the rapidly moving air is itself an erosive agent, the airborne particles are a far more destructive force, dislodging grains as they collide with the soil aggregates.

The way in which the detached soil particles are transported by wind depends mainly on their size. Most of the grains move by saltation, a mechanism that involves short bounces across the soil surface. Particles moving by this method rarely rise more than 30 cm above the ground. Particles with a fine sand size (0.05 to 0.5 mm) are transported particularly easily by saltation. For this reason, soils with fine sandy textures are among the most vulnerable to wind erosion.

The smallest particles detached during erosion, principally silts and clays, may be carried in suspension by wind to great heights and transported thousands of kilometres from their origin. This latter method of particle transport gives rise to dust storms that frequently originate over dry, unprotected soil during windy weather conditions (plate 27a). Despite the dark and ominous appearance of many dust clouds, particles transported in this way generally account for less than 15% of the total amount moved by wind.

Larger grains, up to about 1 mm in size, may move simply by rolling along the soil surface in a process known as soil creep. This is the principal means by which many sand dunes move (plate 24c). In extreme cases of erosion, wind removes all of the finer particles, leaving only a very coarse soil surface known as desert pavement (plate 27b).

Control of wind erosion – Wind erosion can be controlled most effectively by increasing soil moisture and reducing soil surface wind velocity. Moisture greatly increases soil particle cohesiveness, and therefore also increases the wind velocity required to detach individual grains. Where irrigation is available, it is often advisable to moisten the soil surface if dry, windy weather is anticipated. However, few have access to irrigation waters and such quick preventative measures.

Ground-level wind velocity can be reduced by maintaining a well-anchored crop residue, such as stubble mulch from a previous crop. Tillage practices that increase surface roughness, and in this way lower wind velocity and trap soil, have also been used with success. Reduction in wind velocity can be achieved also by planting trees in rows known as shelter-belts (plate 28a). These windbreaks have the additional advantage of providing shelter for farm buildings and a habitat for wildlife.

SOIL SALINITY

Salt is a flavour-enhancing addition to many foods, yet an excess of this substance can kill plants and lay waste to vast stretches of the world's most fertile land. The ancient Romans were well aware of salt's destructive influence on plant growth and used this knowledge for strategic advantage. Following the defeat of Carthage more than 2000 years ago, the Romans spread salt over the land to ensure that this once great society could not re-establish itself. Even today, many of the world's soils are sufficiently degraded by salts that they are unable to support the growth of plants that provide food and fibre for humans, or habitats for wildlife.

Salts that are more soluble than gypsum $(CaSO_4)$ are the most harmful to soil quality. These so-called soluble salts are composed mainly of calcium (Ca^{2+}), magnesium (Mg^{2+}), sodium (Na^+), chloride (Cl^-) and sulphate (SO_4^{2-}). Magnesium sulphate $(MgSO_4)$ and sodium chloride $(NaCl$, table salt$)$ are among the most common soluble salts. These salts originate from mineral weathering, ancient deposits laid down aeons ago, or are added to the soil through rainfall or irrigation. Soils containing an excess of soluble salt are said to be saline. These salt-affected soils occur mainly in arid and semiarid environments, where evaporation exceeds precipitation.

Movement of soluble salts

Saline soils occur most frequently in the lower positions of a landscape, where soluble salts accumulate in what are known as saline seeps (fig 35). These features form when salt-rich groundwater flows above an impermeable layer, towards a lower position in the landscape. Once in the seep area, the salty water moves to the soil surface, where it is evaporated or transpired through plants, thus leaving the salts behind.

During the early stages of salinization, the salts may be visible as white surface crusts near the periphery of the seep area. As salt accumulation increases, the entire surface of the seep area may be covered with a white salt crust (plate 28b) that is visible even from aircraft passing overhead. Soluble salts may accumulate at the soil surface in other, equally visible ways. For example, deeply furrowed soils in saline areas often show a salt crust only on their ridges. Salt-rich water in these soils moves upwards via capillary action to the ridge crests. The water then evaporates at the ridge surface, leaving a salt crust behind (plate 28c).

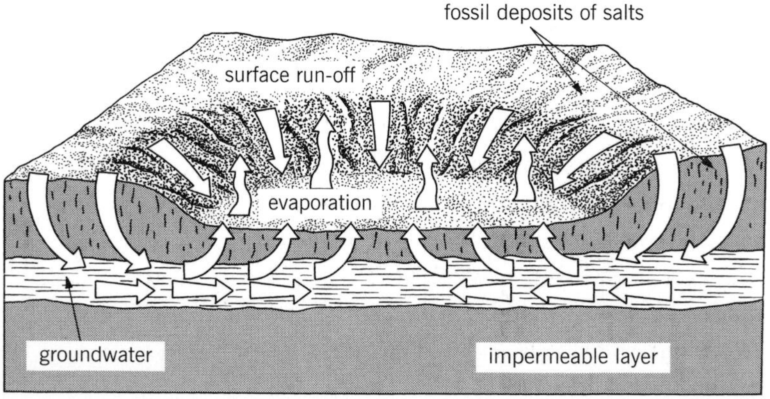

fossil deposits of salts

surface run-off

evaporation

groundwater

impermeable layer

FIG 35
Cross-section showing the accumulation of salts in a depression to produce a saline seep.

PLATE 17A (FAR LEFT)
Hydrangea macrophylla grown in soil of pH greater than 7 are dominated by pink and red flowers.

PLATE 17B (LEFT)
Blue and purple flowers are most abundant on *Hydrangea macrophylla* grown in acidic soils.

PLATE 17C
Poor soil fertility has caused discoloration and reduced size of the geranium leaf (right), as compared to a normal leaf (left).

PLATE 18A (RIGHT)
Nitrogen deficiency in
citrus trees in Iraq,
resulting in yellow leaf
margins (chlorosis).

PLATE 18B (FAR RIGHT)
Large amounts of
nitrogen are lost from rice
paddies through
denitrification.

PLATE 18C
Nitrogen deficiency has
caused this corn leaf to
become chlorotic.

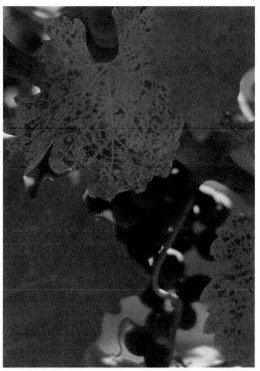

PLATE 19A (FAR LEFT)
Insufficient phosphorus has caused the purple leaf margins of this corn.

PLATE 19B (LEFT)
The pale leaves with purple blotches indicate phosphorus deficiency in these red grapes.

PLATE 19C
Excess phosphorus in lakes and streams may cause algal blooms.

PLATE 20A (RIGHT)
Potassium deficiency in apples appears as necrotic leaf margins.

PLATE 20B (FAR RIGHT)
Necrotic spots on potato leaves caused by potassium deficiency.

PLATE 20C (RIGHT)
Symptoms of severe potassium deficiency in potatoes.

PLATE 20D (FAR RIGHT)
The necrotic spots on these rice plants are symptomatic of potassium deficiency.

PLATE 21A
Lettuce grown in sulphur-deficient soil (left) is dramatically smaller than that which is grown in sulphur-rich soil (right).

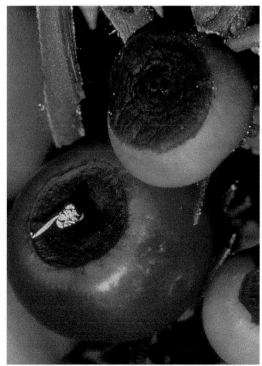

PLATE 21B (FAR LEFT)
Insufficient sulphur results in reduced flowering (foreground) in this field of canola in Saskatchewan, Canada.

PLATE 21C (LEFT)
Calcium deficiency in tomatoes has caused this unsightly scarring of the fruit.

PLATE 22A (RIGHT)
Magnesium deficiency in camellia.

PLATE 22B (FAR RIGHT)
Magnesium deficiency in lilac.

PLATE 22C
Magnesium deficiency in grapefruit causes chlorotic leaves and discoloured fruit.

PLATE 23A (FAR LEFT)
Iron deficiency in this field of soybeans has resulted in the pale green colours visible in the distance.

PLATE 23B (LEFT)
Symptoms of boron deficiency in almonds.

PLATE 23C
Zinc deficiency caused the leaves of this orange tree to be chlorotic.

PLATE 24A (RIGHT)
Compost heap showing various stages of decomposition. Recent, fibrous material at the top grades downwards into darker, humified organic material.

PLATE 24B (FAR RIGHT)
Animal manures provide nutrients and valuable organic material.

PLATE 24C (RIGHT)
Migration of dunes is frequently a major aspect of desertification.

PLATE 24D (FAR RIGHT)
Sheet erosion is accelerated on steep slopes.

PLATE 25A
Extreme sheet erosion in Kenya. Prior to erosion, this boulder rested at the soil surface.

PLATE 25B (FAR LEFT)
Rill erosion.

PLATE 25C (LEFT)
Extreme gully erosion can scar the landscape severely.

PLATE 26A
Crop residues reduce
raindrop impact and
minimize water erosion.

PLATE 26B
Terraced fields, such as
these in Indonesia,
greatly reduce water
erosion.

PLATE 27A (FAR LEFT)
Wind erosion in Mali. The dust storms from such erosion may greatly reduce visibility.

PLATE 27B (LEFT)
Extreme wind erosion removes fine particles to produce desert pavement.

PLATE 27C
These fences in Morocco control wind erosion by reducing ground-level wind velocity.

PLATE 28A
Shelter-belts restrict wind
erosion by reducing
ground-level wind
velocity.

PLATE 28B (RIGHT)
Salt crusts may cover
large areas of highly
saline soils.

PLATE 28C (FAR RIGHT)
Salt crusts on the ridges
of saline soils. Note that
plants are able to grow
on the sides of the ridges,
where the crust is not
present.

PLATE 29A
Necrosis in banana leaf caused by excess soil salinity.

PLATE 29B
Patchy growth of the crop at left is caused by salinity. Note that the crop at right is relatively unaffected.

PLATE 30A (RIGHT)
Heavily contaminated
soils may be unable to
support the growth of
plants.

PLATE 30B (FAR RIGHT)
Industrial emissions may
contain pollutants that
travel great distances
before deposition at the
Earth's surface.

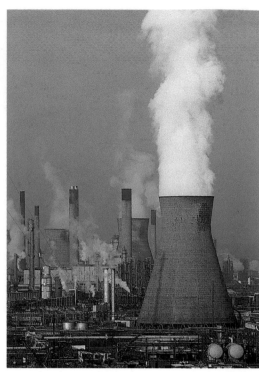

PLATE 30C
Bioremediation of oil-
contaminated shore (left),
compared with the
untreated shore (right).

PLATE 31A
An abundant and sustainable food supply is possible through proper land management.

PLATE 31B
Soils stripped of vegetation, such as these in Niger, are vulnerable to erosion.

PLATE 32A
The prosperity of future generations depends on responsible use of the Earth's resources.

PLATE 32B
Healthy soils support healthy and productive ecosystems.

Measurement of salinity and sodicity

The electrical conductivity (EC) of water increases as salts dissolve and release their electrically charged ions. An indirect measure of soluble salt content in soils can therefore be obtained by measuring the EC of the soil water. Electrical conductivity is expressed in units of decisiemens per metre (dS/m). An EC greater than 4 dS/m indicates a saline soil.

Of all the ions released by soluble salts, sodium (Na^+) is the most problematic. Sodium disperses clay and organic matter, thereby degrading soil structure and reducing macropore space. Consequently, soils high in sodium are poorly aerated and have restricted permeability to water. Sodium content in soils is expressed as either the exchangeable sodium percentage (ESP), or the sodium adsorption ratio (SAR). The ESP indicates the extent to which the cation exchange capacity of a soil is occupied by sodium. Soils with an ESP greater than 15 are known as sodic soils. These soils are considered not saline because their EC is less than 4 dS/m, although their pH values are very high, usually between 8.5 and 10. Soils with an EC greater than 4 dS/m and an ESP greater than 15 are known as saline–sodic soils. These soils have properties of both the saline soils and the sodic soils. However, like the saline soils, saline–sodic soils generally have a favourable structure.

Effects of salinity and sodicity on plants

Soluble salts alter osmotic forces in soils and in this way hinder the uptake of water by plants (plates 29a, b). Because of unfavourable osmotic potentials, a plant may be unable to acquire sufficient water, even if water is abundant. In such cases, a plant is said to experience 'physiological' drought. In the presence of very high salt concentrations water may even reverse its movement, and flow from the plant roots into the soil solution.

Plants vary in their sensitivity to salty soils. Those with the lowest salt tolerance include tomatoes, onions and lettuce. At the other extreme are the halophytes, which occur most frequently in salt marshes, beaches and other saline environments. One of the most noteworthy halophytes is the red saltwort (*Salicornia rubra*), a highly salt-tolerant plant that commonly occurs near saline seeps.

The deleterious effects of salts on plants are caused not only by osmotic forces, but also by toxic levels of sodium and chloride. Fruit crops and woody ornamentals are especially sensitive to high levels of these elements. Moreover, the high pH caused by excess sodium may result in micronutrient deficiencies.

It is common practice in many cities of temperate regions to use salts during winter months as de-icing agents on roads and pedestrian areas. Heavy applications of these salts frequently damage nearby vegetation. With sufficient precipitation, however, these salts are leached out of the soil and the vegetation can recover. Although salts are still frequently used, the undesirable effects of excess sodium have prompted many municipalities to now use KCl rather than NaCl.

Reclamation of salt-affected soils

The reclamation of salt-affected soils can be difficult and lengthy. Despite the challenges inherent in such a difficult task, one can improve the chances for successful remediation by following three rules. The first step is to establish good internal drainage to facilitate the leaching of salts. Many salt-affected soils currently have adequate drainage in place; others require the implementation of artificial drainage systems, or the addition of deep-rooted plants to lower the water table. Second, excess sodium in sodic and saline–sodic soils must be replaced by calcium. This is achieved most effectively and economically with the addition of gypsum. And third, the salts must be leached from the soil. This can be accomplished with the liberal application of good quality irrigation water. Where irrigation water is not available, natural precipitation must be relied upon to leach the salts.

POLLUTION

Soils have been the recipient of society's waste throughout human history. Since the

---

SALINITY 'DOWN UNDER'

Soils in the coastal regions of Australia have for aeons been receiving small but significant amounts of salt from the sea-spray that drifts inland. As natural precipitation infiltrated down-wards through the soil, the salts were carried downwards also, accumulating at a depth far below the root zone of most plants. The robust and deep-rooting eucalyptus and acacia trees of the region extracted water from great depths and in this way ensured that the water and salt did not migrate upwards.

Settlement of Australia by Europeans had begun in earnest by the middle of the 19th century. These early settlers were drawn in great numbers to the fertile but forested lands of south-west Australia. Large areas of indigenous forest in this region were cleared in response to the growing demand for food. Replacement of the native trees with much smaller, shallow-rooted crops and grasses greatly reduced the vegetative demand for water. Consequently, a greater proportion of the annual precipitation remained in the soil where it ultimately collected at some depth below the surface. As the amount of sub-surface water increased with each passing year, the salts that had accumulated underground in previous geological time soon began to dissolve. As decades passed, the waters grew more saline and began to rise.

The early settlers to Australia could not have foreseen the unfortunate consequences of their well-intentioned actions. As farmers continued to reap the benefits of the fertile soil, the unseen salty waters continued their relentless upwards rise to the soil surface. By the middle of the 20th century, denuded soil and salt crusts began appearing in low-lying areas of fields. Soon these barren spots had expanded to cover large areas of the landscape. Today, nearly 85 million hectares of Australia's soils have been degraded by salts.

Unfortunately, Australia is not a unique case. Human activities have contributed to the saliniza-tion of soils in other parts of the world throughout our history. Irrigated lands are particularly susceptible to salinization. Reclamation of salt-affected soils in Australia and elsewhere can be difficult and long-term. The benefits, however, can be immeasurable.

Industrial Revolution, however, the amount, toxicity and persistence of this waste have increased significantly. As these myriad contaminants enter a soil they quickly become enmeshed in its chemical and biological cycles. Some pollutants occur in soils in forms that are relatively unavailable and benign. Other pollutants rapidly enter the food chain and disrupt the normal function of biological communities or pose a threat to human health (plate 30a).

Inorganic contaminants

The main inorganic soil contaminants are arsenic, cadmium, chromium, copper, lead, mercury, molybdenum, nickel, selenium and zinc. Of these, arsenic, cadmium, lead and mercury are the most toxic to humans. These inorganic contaminants have many sources, including smelting, electroplating, fossil fuel combustion and many other industry-related processes (plate 30b). Once released, these contaminants may be carried great distances by wind or water before they are deposited on soil.

The specific sources of the most toxic inorganic contaminants deserve mention. Lead, a potent neurotoxin, was for many years derived from vehicle emissions and certain paints. Arsenic, a once-popular antibiotic, may originate from mine tailings, animal feed supplements or pesticides. For many years arsenic-containing insecticides were used to control pests on cotton, lawns and fruit crops. Consequently, this element frequently accumulated to toxic levels in the associated soils. Cadmium, which can accumulate in animal tissues and contribute to kidney damage in humans, may be added to soil as an impurity in phosphate fertilizers.

Sewage sludges can also contain significant amounts of many inorganic pollutants. For this reason, the composition and quantity of municipal sludges applied to agricultural land is strictly regulated.

Elements may occur in radioactive forms known as radionuclides. Some radionuclides occur naturally, others have been introduced to the environment by human activities such as nuclear weapons testing and nuclear power generation. Following the failure in 1986 of the nuclear power station at Chernobyl, Ukraine, radionuclides of strontium (^{90}Sr), iodine (^{131}I) and caesium (^{137}Cs) were deposited on many soils across Europe. Ten years after this accident, the incidence of thyroid cancers in people living near Chernobyl was found to be nearly 20 times greater than in other populations.

Mobility and bioavailability – Cadmium, copper, nickel and zinc occur in soils generally as cations that bind tightly to the negatively charged minerals and organic matter. At pH values greater than 6.5 these metals are only slowly available to plants, but they become considerably more labile under acidic conditions. Lead is highly immobile in most soils and generally unavailable to plants, except at low pH. The principal threat to human health from lead-polluted soils is the risk of ingesting contaminated soil particles.

Mercury in well-aerated soils is generally bound to minerals and organic matter and in this form it is largely unavailable to plants. However, in poorly drained soils and swampy areas, micro-organisms can synthesize organo-mercury compounds that are both highly toxic and readily absorbed by plants and animals. These organic forms of mercury may accumulate in fish and other aquatic animals to levels that are toxic to humans.

Arsenic, molybdenum and selenium are usually present in soils as anions. Consequently, their mobility and bio-availability are lowest under acidic soil conditions, when the amount of positive charge on minerals and organic matter is greatest. Chromium, like arsenic and selenium, may occur in different oxidation states depending on how well a soil is aerated. The oxidized form of chromium, chromate, is both highly mobile and very toxic. Fortunately, a small amount of organic matter is sufficient to convert chromate to the benign form.

The radionuclides ^{90}Sr and ^{137}Cs behave in soils similarly to calcium and potassium, respectively. Caesium 137 can be fixed tightly by soil minerals such as micas and vermiculites, thereby reducing its availability significantly. Strontium 90 is perhaps more of a concern than ^{137}Cs because the former may enter the food supply by substituting for calcium and thus accumulate to considerable concentrations in the bones of humans and other animals. As a general rule, the mobility and bioavailability of radionuclides and other inorganic contaminants will be greatest in acidic, sandy soils with low cation exchange capacity and little organic matter.

Remediation – The first and most important consideration when remediating soils contaminated with inorganic pollutants is to curtail the addition of further toxins. Second, the contaminants should be immobilized to limit their uptake by plants or leaching to groundwater. For most contaminants this can be achieved by adding lime to increase the pH to neutral or higher. A third way to mitigate metal pollution of soils is to grow plants that hyperaccumulate the metal of concern. Certain plants, such as those of the genus *Thlaspi* (from the Cruciferae family), can accumulate zinc with such efficiency that their tissues contain 4% of this metal. In some cases the harvested plants may contain enough zinc to be used for smelting new metal.

Organic contaminants

Thousands of synthetic organic compounds are introduced to the environment each year. These organic substances include pesticides, oils, solvents and other hydrocarbons. Pesticides, the most important group of organic pollutants, can be divided into three classes depending on their target organism: herbicides, insecticides and fungicides. Pesticides vary greatly with respect to their persistence and toxicity. Chlorinated compounds, some of which may persist in soils for 15 years or more, are the most recalcitrant. Organo-phosphate pesticides generally biodegrade much more quickly, but they can be highly toxic to humans. The most desirable pesticides are those that degrade rapidly and whose toxicity is highly specific.

Fate of organic contaminants – Most organic contaminants, regardless of their origin, ultimately reach the soil, where they are then subjected to a complex web of biogeochemical reactions (fig 36). Many contaminants are rapidly adsorbed by soil minerals and organic matter. The extent of adsorption increases with the amount of organic matter and clay. Larger and more highly charged contaminant molecules are

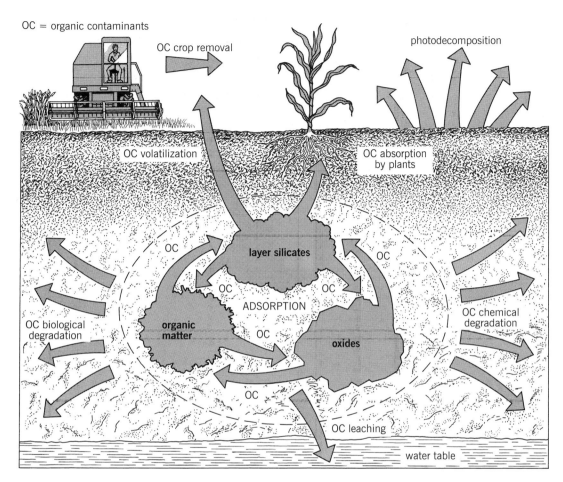

OC = organic contaminants

OC crop removal

photodecomposition

OC volatilization

OC absorption by plants

OC

layer silicates

OC

OC

OC

ADSORPTION

organic matter

OC

oxides

OC biological degradation

OC chemical degradation

OC

OC leaching

water table

FIG 36
Organic contaminants (OC) are subjected to many reactions in soil.

also adsorbed more strongly by soils. As a general rule, pesticides lose their efficacy once they become adsorbed. Organic contaminants that are weakly adsorbed and highly water soluble may be leached easily from the soil profile. The potential for leaching is greatest in highly permeable sandy soils with little cation exchange capacity.

Organic contaminants in soils may also be degraded to simpler, less toxic forms. Three mechanisms of degradation are recognized: biodegradation, chemical degradation and photodecomposition. Of these, biological degradation is the most

important. The rapidly growing technology known as bioremediation employs biological degradation to detoxify contaminants and thus improve soil quality (see box, page 102).

Contaminants may be lost from the soil through volatilization. Compounds that are highly volatile, such as solvents, fuels and some pesticides, may be lost in considerable amounts in this way. Pesticides that are absorbed by plants may also be lost from the soil when crops are harvested. Regulatory bodies in many countries conduct exhaustive tests to ensure that such pesticide residues in plants are safe for human consumption.

BIOREMEDIATION – NATURE'S CHIMNEY-SWEEP

Bioremediation is a process that incorporates plants, micro-organisms or enzymes to degrade contaminants and improve soil and ecosystem health. This process is often facilitated with the addition of nutrients or the introduction of organisms that are specially designed to degrade the contaminant of interest. Bioremediation was used with considerable success to restore the polluted shoreline of Alaska following the spill in 1989 of crude oil from the *Exxon Valdez* oil tanker (plate 30c). Composting, it should be noted, is also a form of bioremediation, as the indigenous organisms in the compost heap can effectively degrade resident contaminants.

Phytoremediation is the form of bioremediation that uses plants to restore soil quality. Direct phytoremediation refers to the assimilation of organic contaminants and metals directly by plant roots. Indirect phytoremediation is the process whereby root exudates stimulate rhizosphere micro-organisms to degrade nearby organic contaminants. Recent experiments have shown that rhizosphere microbial populations are able to degrade certain organic contaminants (e.g. trichloroethylene, petroleum hydrocarbons) nearly twice as fast as the non-rhizosphere microbial populations. Clearly, the complexity and richness of rhizosphere interactions benefit not only the plant and associated micro-organisms but also the greater soil environment.

A number of constraints hamper the wider application of bioremediation strategies. Environments that are too cold, dry, nutrient poor, or in which the contaminant is not bioavailable, limit the suitability of bioremediation as a method of environmental restoration. Despite these limitations, bioremediation remains one of the most promising strategies for the restoration of contaminated environments.

SOILS AND FOOD SUPPLY

Of the world's total land area, only about 25% is suitable for cultivation. The remaining 75% is either swampland, mountainous, too cold or dry, or fit only for animal grazing. The relatively small area that is suitable for cultivation has been placed under increasing demands as human populations have grown since the beginning of agriculture nearly 10,000 years ago.

During the last century, advances in medical science drastically reduced death rates, especially among children, and led to an unprecedented population growth. In light of this explosive growth in human populations, and the limited land area available for food production, many social scientists predicted widespread starvation. However, these scientists did not foresee the development of intensive agriculture, and the consequent increase in food production during this same period.

Intensive agriculture and food production

Between 1960 and 1990, food production grew faster than population in all areas of the world except sub-Saharan Africa (fig 37). This increase was the result of greater

FAMINE IN AFRICA

In 1967, annual grain production in Africa peaked at about 180 kg per person, an amount considered barely adequate to sustain human life. Two decades later, this production had fallen to 120 kg per person. Africa is the only region of the world to have experienced such a decrease in per capita food production. As a consequence, starvation and despair in this continent have become all too common.

There is a perception among many that the frequent famines in Africa are caused only by drought. In reality, the plight of Africa can be attributed to many factors: large areas of very fragile soils, various political, social and economic forces, as well as severe and frequent droughts. All of these factors have contributed to the mismanagement and degradation of the continent's soils and vegetation.

Most soils in Africa are highly weathered, nutrient poor and often suffer from aluminium toxicity. As new land was brought into agricultural use in an effort to increase food production, the native forests were cleared, thus robbing the soil of its protective cover (plate 31b). These trees that once served as nutrient pumps, bringing essential elements from great depths back to the soil surface, were now being burned for heating and cooking. With sustained nutrient losses caused by crop removal, leaching and surface run-off, the fertility of these marginal soils declined still further. As firewood disappeared along with the forests, families began to use crop residues and animal dung for cooking and heating. Consequently, these valuable resources were not returned to the soil where they would have provided much-needed nutrients and organic material. As organic matter was lost, soil structure deteriorated, with a consequent increase in erosion. The continued decline in soil fertility resulted in decreased yields of protein-rich grains and legumes. In an effort to maintain daily caloric intake, farmers began to grow energy-rich, but less nutritious crops, such as cassava, yams and potatoes.

The first step to end the vicious cycle of soil and ecosystem degradation is to minimize the burning of vegetation and restrict the cultivation of marginal soils. Grazing and cultivation should be permitted only on stable soils. These measures will increase the amount of organic material returned to the soil, thereby improving soil nutrient status and structure. Better fertility and structure, in turn, improves plant growth and water infiltration, leading ultimately to reduced soil erosion.

Various trees and agricultural crops can be grown together in what are known as agroforestry systems. The fast-growing, nitrogen-fixing trees in these systems provide additional nitrogen, act as nutrient pumps and protect the soil from erosion. It must be emphasized that efforts to improve soil quality can succeed only if local farmers are given sufficient incentive to implement and maintain these measures.

nutrient additions, the cultivation of high-yielding crops and the implementation of comprehensive water management strategies, including irrigation. In Asia, where yield increases were particularly impressive, the term Green Revolution has been used to describe this increase in food production.

Intensive agriculture has brought about several important benefits. Most importantly, increased grain supplies reduced starvation and malnutrition in all parts of the world (plate 31a). Greater yields have also limited the amount of land needed for food production. For example, because of global crop intensification, an area nearly twice the size of Argentina has been saved from cultivation. Also, greater fertilizer additions have increased the amounts of nitrogen, phosphorus and potassium in many soils.

Unfortunately, intensive agriculture has been detrimental to some soils and ecosystems. Excess nutrient additions in many areas of the world, particularly Western Europe, have resulted in eutrophication and nitrate contamination of groundwater. Also, irrigation with poor quality water has led to salinization of many soils in arid regions. Furthermore, the growth of monoculture crops over large areas for many seasons has reduced soil biodiversity and increased plant diseases. Such intensive farming systems are essential if food is to be produced in amounts sufficient to feed the world's growing population. However, soil quality must be maintained to ensure that these high yields can be continued indefinitely.

FIG 37
Since 1961, per capita food production has increased in all parts of the world except sub-Saharan Africa.

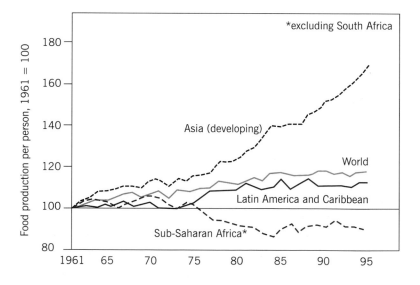

WHAT LIES AHEAD?

As the 21st century begins, the world's six billion inhabitants are placing unprecedented demands on the global soil resource. As the human family continues to grow both in numbers and prosperity, so too will the demand for food and fibre. For example, the global migration of people from towns and villages to more prosperous lifestyles in cities is often accompanied by changes in diet. Among the most important dietary changes is an increase in the amount of animal products consumed. Given that approximately eight kilograms of grain are required to produce one kilogram of beef, such changes in diet on a large scale can very quickly increase the demand for food.

Healthy soils are essential not only for the provision of food and fibre but also for the normal functioning of ecosystems, the preservation of biodiversity and the maintenance of recreational areas. As we attempt to satisfy our growing material needs, we will be confronted with the ever-present risk of soil degradation caused by compaction, salinization, erosion, pollution, desertification and loss of soil biodiversity. Moreover, a changing global climate is likely to have a significant influence on the world's soils. Although the effects of global warming on soils cannot at present be determined with certainty, there is likely to be an increased risk of desertification, and loss of coastal lands as a consequence of rising sea levels.

Future soil management strategies must be designed to meet two essential requirements: (i) the need for increased food production and (ii) the need to maintain soil and environmental quality. These dual requirements are most likely to be satisfied by adopting a food production system that minimizes disruption to the natural ecosystem. Such systems would, by virtue of the need to increase food production, employ intensive agriculture practices. However, it is now recognized that greater fertilizer additions will not always produce greater yields. The nutrient levels of many soils, with the exception of those in sub-Saharan Africa and other developing regions, are already near optimal levels. Rather, yield increases are more likely to arise from advances in nutrient and water use efficiency, improved crop rotations and the utilization of higher-yielding crop varieties.

There will be many challenges to overcome, but there is every reason for optimism. Policymakers and the general public are becoming increasingly cognisant of the importance and fragility of the world's soils. One of the most difficult challenges will be to implement practical and effective solutions to food production and land management in developing countries. The primary consideration of people in many of these regions is, understandably, not soil preservation, but self preservation. Those in developed countries must ensure that the less affluent among us have the resources to develop sustainable agroecosystems as we work together to improve our common future. It is indeed a global challenge to protect for future generations this fragile resource, the skin of the Earth.

Glossary

acid sulphate soils – soils that are highly acidic because of large amounts of sulphur that have been oxidized to sulphuric acid

actinomycetes – a group of organisms related to both fungi and bacteria

adsorption – process by which ions and molecules in solution are bound to the surface of solids

aeolian parent material – earth material transported and deposited by wind

aggregation – natural process whereby primary soil particles are bound together by physical, chemical and biological forces to form aggregates

algal bloom – rapid increase in algae growth of surface waters, caused by increased supply of nutrients, especially phosphorus

allophane – poorly crystalline clay-sized mineral that occurs mainly in soils formed from volcanic ash

alluvium – eroded sediments deposited by flowing water, particularly rivers and streams

amendment, soil – any substance, such as lime, manure, and inorganic fertilizer, that is added to soil to improve productivity

ammonium fixation – the binding of ammonium ions by layer silicates with such strength that these ions are not available to plants; similar fixation may occur for potassium

anion – negatively charged ion

available nutrients – that portion of soil nutrient that is in a form readily available to plants

available water – the portion of soil water that can be readily absorbed by plant roots; available water exists between field capacity and permanent wilting point

bioremediation – the use of biological agents, such as plants and micro-organisms, to decontaminate and restore polluted soils

buffering capacity – ability of a soil to resist a change in pH; this ability is related mainly to the amount of humus and clay

bulk density – the mass of dry soil per unit of bulk volume, which includes both solids and air space

buried soil – soil covered by a layer of alluvial or loessal material

cation – positively charged ion

cation exchange capacity (CEC) – the sum of exchangeable cations that may be held by a soil

chelates – stable chemical associations between organic molecules and metals; chelating agents are sometimes added to soil to increase the availability of trace elements

chlorosis – condition in plants characterized by pale yellow and green leaves

clay minerals – naturally occurring materials, principally layer silicates, that occur within the clay size (< 0.002 mm) fraction

clod – a compact and discrete mass of soil that is produced artificially, as during ploughing or digging

colluvium – unsorted, unconsolidated earth material deposited by gravity at the base of slopes

compost – organic residues that have been modified by microbial decomposition to produce a material that may be used as a soil amendment or potting mix

crop rotation – a planned sequence of crops that are grown on the same area of land; such rotations serve to minimize disease and maximize nutrient use efficiency

cryoturbation – physical churning of soil caused by repeated freezing and thawing

denitrification – transformation of nitrate (or nitrite) to gaseous forms of nitrogen

desert pavement – layer of closely packed pebbles, boulders and rock fragments that remain after the finer particles have been removed by wind

electrical conductivity (EC) – capacity of soil water to conduct electricity; used to measure the soluble salt content of soil

eluviation – removal from a soil layer of material in solution or suspension

essential elements – chemical elements required for the normal growth and reproduction of plants

eutrophication – nutrient enrichment of ponds, lakes and streams; the subsequent growth of aquatic organisms leads ultimately to oxygen depletion

exchangeable sodium percentage (ESP) – the percentage of a soil's cation exchange capacity that is occupied by exchangeable sodium

field capacity – the amount of water remaining in a well-drained soil two or three days after its having been saturated

fluvial deposits – earth material transported and deposited by rivers or streams

fungi – diverse group of single-celled (e.g. yeasts) and multicellular (e.g. moulds and mushrooms) organisms

gibbsite – an aluminium hydroxide mineral that is most abundant in highly weathered soils, such as Ultisols and Oxisols

glacial till – unsorted and unstratified earth material that has been deposited by glacial ice and which consists of any proportion of sand, silt, clay and boulders

goethite – an iron oxide mineral that occurs in almost every soil type and climatic region and which is responsible for the yellow-brown colour of many soils

green manure – plant material, frequently legumes, incorporated into soil while green to improve soil quality

halophyte – a plant that requires, or can tolerate, a saline environment

heavy metals – those metals, frequently toxic, with densities greater than 5 g/cm^3; the most common heavy metals in soils include Cd, Co, Cr, Cu, Fe, Hg, Mn, Mo, Ni, Pb and Zn

horizon, soil – a layer of soil material, approximately parallel to the Earth's surface, that differs from adjacent layers in various physical, chemical, and biological properties and characteristics

humus – the highly decomposed and relatively stable fraction of the soil organic matter

illuviation – deposition of soil material from an upper to a lower horizon in the soil profile

imogolite – poorly crystalline clay-sized mineral that occurs mainly in soils formed from volcanic ash

ions – atoms or molecules that are electrically charged

kaolinite – a layer silicate mineral that is most abundant in highly weathered soils

lacustrine deposit – material deposited by lake water which, when exposed, may serve as parent material for lacustrine soils

lime – various acid-neutralizing materials that are added to soil to raise pH and improve soil conditions

limiting factor – any factor (e.g. nutrient, sunlight, heat) which limits the growth and reproduction of a plant

loam – a textural class which contains a moderate amount of sand, silt and clay

loess – earth material, composed mainly of silt-sized particles, that has been transported and deposited by wind

macronutrients – essential elements required in relatively large amounts (i.e. C, H, O, N, P, K, S, Ca, Mg)

micronutrients – essential elements required in small amounts (i.e. Fe, Mn, B, Zn, Cu, Cl, Mo, Co, Ni)

mineralization – transformation of an element from an organic form to an inorganic form as a consequence of microbial activity

mineral soil – a soil whose composition and properties are dominated by minerals; mineral soils usually contain less than 20% organic matter

montmorillonite – a layer silicate mineral that expands when wetted and which may dominate moderately weathered soils of temperate regions

mottles – spots and blotches, usually rust-coloured, that indicate periodic wetting and drying

mulch – any material, such as leaves, sawdust and straw, that is spread over the soil surface as protection against erosion, crusting, or freezing

mycorrhizae – structure formed from the symbiotic association between filamentous fungi and the roots of higher plants; these associations increase nutrient uptake by plants

necrosis – death of plant tissues as indicated by their discoloration and dehydration

nematodes – very small worms that are important in soils because they may attack and damage plant roots

nitrification – transformation of ammonium (NH_4^+) to nitrate (NO_3^-) by bacteria

nitrogen fixation – conversion of elemental nitrogen (N_2) to organic forms that may be used in biological processes

organic soil – a soil in which more than one-half of the profile is composed of organic materials

palaeosol – a soil formed on a landscape in the past and which has distinctive morphological features resulting from a soil-forming environment that no longer exists

parent material – the unconsolidated mineral or organic material from which a soil forms

ped – a unit of soil structure (e.g. granule, plate, block, or column) formed by natural processes

permafrost – permanently frozen layer within or below the soil

permanent wilting point – the soil moisture content at which plants wilt and subsequently fail to recover when placed in a humid atmosphere

pH, soil – a measure of the acidity (or alkalinity) of a soil

pore size distribution – the volume fractions of the various size ranges of pores in a soil

profile, soil – a vertical section through all the horizons of a soil, extending into the parent material

protozoa – single-celled organisms that prey on bacteria, fungi and other soil microbes

regolith – unconsolidate material at the Earth's surface comprising both soil and weathered rock

residual material – unconsolidated rock and mineral material formed by the disintegration of solid rock *in situ*

rhizosphere – the zone of soil in the immediate vicinity of plant roots in which microbial populations have been altered by the presence and activity of roots

saline seep – an area of land in which saline water moves to the soil surface and evaporates, producing a region of high salt concentration

saline soil – a non-sodic soil which contains sufficient soluble salts to impair plant growth

sewage sludge (biosolids) – solids and various dissolved materials removed from sewage during wastewater treatment

shelterbelt – a barrier of trees that reduces wind velocity and therefore decreases the erosive forces of wind

sodic soil – a soil that contains sufficient exchangeable sodium to adversely affect soil structure and plant growth

soil classification – the systematic arrangement of soils into groups and categories on the basis of physical, chemical and biological characteristics

soil conservation – implementation of soil management and land use strategies that serve to protect the soil from erosion, nutrient depletion, and deterioration by natural or anthropogenic factors

soil erosion – the wearing away of the land surface by various natural or anthropogenic agents, including wind, water, ice, and tillage

soil fertility – the quality of a soil that enables it to provide nutrients in adequate amounts and in the proper balance to optimize plant growth

soil organic matter – plant and animal residues at all stages of decomposition, as well as cells, tissues and organic substances produced by soil organisms; the organic fraction is sometimes determined to be that soil organic material which passes through a 2 mm sieve

soil quality – the capacity of a soil to sustain biological productivity, maintain environmental quality, and promote plant and animal health

soil separate – any of the three mineral particle sizes (i.e. sand, silt, or clay)

soil structure – the combination or arrangement of primary soil particles into larger units, or peds

soil texture – the relative proportions of the various soil separates

solum – the upper, most highly weathered part of a soil profile (i.e. the A, E, and B horizons)

xenobiotic – a compound foreign to biological systems; often refers to anthropogenic compounds that are resistant to decomposition

Index

Italics refer to captions, illustrations and tables. Bold refers to feature boxes. Where FAO-UNESCO classification is given, *Soil Taxonomy* equivalent is given in parentheses. See p.28 for comprehensive table.

Further information

FURTHER READING

The Chemistry of Soils, G. Sposito. Oxford University Press, Oxford, 1989.
[presents the fundamental concepts of soil chemistry]

Environmental Soil Chemistry, D.L. Sparks. Academic Press, London, 1995.
[describes the environmentally important reactions in soil, with emphasis on those that govern the fate of pollutants]

The Nature and Properties of Soils, N.C. Brady and R.R. Weil. Prentice Hall, London, 1999.
[covers the biological, chemical, and physical properties of soils, including their environmental applications and management]

Out of the Earth – Civilization and the Life of the Soil, D. Hillel. University of California Press, Berkeley, 1991.
[explains with many examples the role of soil in the evolution of civilizations]

Principles and Applications of Soil Microbiology, D.M. Sylvia, J.J. Fuhrmann, P.G. Hartel and D.A. Zuberer. Prentice Hall, London, 1999.
[provides a comprehensive introduction to the rapidly changing discipline of soil microbiology]

Principles of Plant Nutrition, K. Mengel and E.A. Kirkby. International Potash Institute, Bern, 1987.
[a comprehensive examination of plant nutrition]

Soil Conditions and Plant Growth, E.W. Russell. Wiley, 1988.
[presents virtually all aspects of soils related to plant growth]

Soil Fertility and Fertilizers: An Introduction to Nutrient Management, J. Havlin, S.L. Tisdale and J.D. Beaton. Prentice Hall, London, 1998.
[provides a thorough coverage of soil fertility, plant nutrition and nutrient management]

Soils and Environment, S. Ellis and A. Mellor. Routledge Press, London, 1995.
[examines the many ways in which soils influence, and are influenced by, the environment]

Soils in Our Environment, R.W. Miller and D.T. Gardiner. Prentice Hall, London, 1998.
[thorough introduction to soils and their management]

INTERNET RESOURCES

NB. Web site addresses are subject to change.

Australian Soil Science Society
http://www.waite.adelaide.edu.au/ASSSISA/

British Society of Soil Science
http://www.bsss.bangor.ac.uk/
[links to many national and international sites]

CSIRO Land and Water
http://www.clw.csiro.au/

Food and Agriculture Organisation
http://www.fao.org/
[links to many projects, funded by the United Nations, whose aim is to improve soil management in developing countries]

Garden Soil Management
http://www.ag.ohiostate.edu/~ohioline/lines/hygs.html#SOILS
[includes a comprehensive list of fact sheets related to the management of garden soils]

International Soil Reference and Information Centre
http://www.isric.nl/

National Soil Survey Centre
http://www.statlab.iastate.edu/soils/nssc/
[links to soil maps and other soil survey data for the United States]

Soil Science Education Pages
http://ltpwww.gsfc.nasa.gov/globe/index.htm
[a fun and educational site designed for primary school children and teachers, with many links to similar sites]
http://www.nhq.nrcs.usda.gov/CCS/squirm/skworm.html
[excellent introduction to soils for primary school children]
http://www.fieldmuseum.org/ua/
[a soil education site which includes a virtual tour of soil]

Soil Science Society of America
http://www.soils.org/

USDA Natural Resources Conservation Service
http://www.nrcs.usda.gov/
[official site of the US Department of Agriculture's Natural Resources Conservation Service]

Virtual Soil Science Library
http://www.metla.fi/info/vlib/soils/
[a comprehensive library of soil science resources]

World Soil Resources
http://www.nhq.nrcs.usda.gov/WSR/
[links to international soil resources]

PICTURE CREDITS

PLATES

J. Akworth FRR, 31a; Laurie Campbell, 30b, 32b; Centre Technique de Coopération Agricole et Rurale, 24c; Chris Mack and Oliver Chadwick, 28c, 29b; Craig Ross – NZSSS, 10b; Bill Dubbin (NHM), 2a, 3a; Food and Agriculture Organization of the United Nations, 26b 27c; Dr E. A. Fitzpatrick, 25b, 28b; Stephen M. Hinton, ExxonMobil Corporate Strategic Research Company, Annandale, New Jersey, 30c; David Hoffman, 30a; ISRIC, 1a-c, 3c, 4a, c, d, 5a-d, 6a, b, 7b, 8d, 9b-d, 10a, c, 11b-d, 12a, b, 13a-d, 14b, 15b, c, 18a, b, 24b, 25a, 26a, 27b, 28a, 29a, 31b; Dave John (NHM), 19c; Koninklijk Institute voor de Tropen, 6d, 27a; Toni Lawson-Hall – Windermere, 17a, 17b; Ohio State University, 8b; Potash & Phosphate Institute, 18c, 19a, b, 20a-d, 21a-c, 22c, 23a-c; Tim Sandall, The Garden, 17c, 22a, b; J. A. Stanturf, 8a; Dr M. P. Searle, 9a; UNEP, 6c, 7a; UNESCO, 3b, 32a; United States Department of Agriculture, Natural Resources Conservation Service Soil Survey Division, World Soil Resources (1999), 16; USAID, 15a, d; Tony Waltham, front cover, 2b, 4b, 10d, 11a, 14a, 24d, 25c.

FIGURES

John Darbyshire, 20; Prof. Dindal, 17; Mike Eaton, 1, 3, 11, 13-15, 25, 29, 35, 36; Food and Agriculture Organization of the United Nations, 34; Prof. Dr H. Graf von Reichenbach, 5; ISRIC, 18; Judith John, 22; Mercer Design, 2, 4, 9, 12, 16, 24, 26-28, 32, 33, 37; Craig Ross – NZSSS, 8.

ACKNOWLEDGEMENTS

The author thanks Andy Fleet, Frank Krell, William Purvis, Chris Stanley, Alan Warren (The Natural History Museum, London), Janet Cotter-Howells (University of Exeter), Harvey Doner (University of California, Berkeley), and Ahmet Mermut (University of Saskatchewan), each of whom provided valuable comments to improve the clarity and accuracy of the text.

Thanks also to Albert Bos, Dick Creutzberg, and Otto Spaargaren at the International Soil Reference and Information Centre, Wageningen, The Netherlands, for assistance with gathering many of the images.